PROGRESS

Nelson Mathematics

1

for Cambridge International A Level

Pure Mathematics

L. Bostock • S. Chandler • T. Jennings

Oxford excellence for Cambridge A Level

OXFORD

Great Clarendon Street, Oxford, OX2 6DP, United Kingdom

Oxford University Press is a department of the University of Oxford.
It furthers the University's objective of excellence in research, scholarship,
and education by publishing worldwide. Oxford is a registered trade mark of
Oxford University Press in the UK and in certain other countries

British Library Cataloguing in Publication Data
Data available

978-1-4085-1558-7

10 9 8 7 6 5 4 3 2

MIX
Paper from
responsible sources
FSC
www.fsc.org   FSC® C007785

Printed in Great Britain by Bell and Bain Ltd., Glasgow

Acknowledgements

Cover photograph: Jim Zuckerman/Alamy
Illustrations: Tech-Set Ltd, Gateshead
Page make-up: Tech-Set Ltd, Gateshead

Previous exam questions from Cambridge International AS and A Level Mathematics 9709 reproduced by
permission of the University of Cambridge Local Examinations Syndicate:

Page 34 Q1: paper 1 question 5 November 2009, Q2: paper 11 question 7 November 2007, Q3: paper 13
question 7 June 2010, Q4: paper 13 question 8 June 2010, Q5: paper 1 question 6 November 2007; Page 35
Q6: paper 11 question 8 June 2010, Q7: paper 1 question 5 June 2009, Q8: paper 1 question 8 June 2009, Q9:
paper 1 question 5 part one June 2008, Q10: paper 1 question 11 June 200; Page 36 Q11: paper 1 question
5 June 2007, Q12: paper 1 question 5 June 2004, Q13: paper 1 question 6 June 2004; Page 77 Q1: paper 1
question 10 November 2008, Q2: paper 11 question 7 November 2009, Q3: paper 1 question 11 November
2007, Q4: paper 11 question 7 June 2010, Q5: paper 11 question 9 June 2010, Q6: paper 1 question 10 June
2009; Page 78 Q7: paper 1 question 4 June 2008, Q8: paper 1 question 6 June 2008, Q9: paper 1 question 11
June 2007, Q10: paper 1 question 8 June 2004, Q11: paper 1 question 10 June 2004, Q12: paper 1 question 1
November 2007, Q13: paper 1 question 2 June 2009, Q14: paper 1 question 1 June 2007; Page 141 Q1: paper
1 question 1 November 2008, Q2: paper 1 question 2 November 2008, Q3: paper 1 question 3 November
2008, Q4: paper 1 question 4 November 2008, Q5: paper 11 question 1 November 2009, Q6: paper 11
question 2 November 2009, Q7: paper 1 question 10 November 2007, Q8: paper 13 question 9 June 2010,
Q9: paper 1 question 7 June 2009, Page 142 Q10: paper 1 question 2 June 2008, Q11: paper 1 question 8 June
2007, Q12: paper 1 question 2 June 2007, Q13: paper 1 question 3 June 2007, Q14: paper 1 question 9 June
2004

The University of Cambridge Local Examinations Syndicate bears no responsibility for the example answers
to questions taken from its past question papers which are contained in this publication.

P1 Contents

Introduction	**vi**

Chapter 1 Quadratic equations — **1**

- Expressions and equations — 1
- Quadratic equations — 1
- Solution by factorising — 1
- Completing the square — 3
- The formula for solving a quadratic equation — 5
- Equations that are quadratic in a function of x — 6
- Properties of the roots of a quadratic equation — 6
- Solution of one linear and one quadratic equation — 9

Chapter 2 Coordinate geometry 1 — **11**

- Cartesian coordinates — 11
- The length of a line joining two points — 11
- The midpoint of a line joining two points — 12
- Gradient — 13
- Parallel lines — 15
- Perpendicular lines — 15
- Mixed problems — 16

Chapter 3 Coordinate geometry 2 — **18**

- The equation of a straight line — 18
- Finding the equation of a straight line — 20
- Intersection — 23

Chapter 4 Circular measure — **25**

- The radian — 25
- The length of an arc — 27
- The area of a sector — 28
- Mixed problems — 30

Summary 1 — **33**

Chapter 5 Functions — **37**

- Mappings — 37
- Functions — 38
- Domain and range — 38
- Curve sketching — 40
- Inverse functions — 44
- The graph of a function and its inverse — 44
- Composition of functions — 46

Chapter 6 Inequalities and intersection of curves   49

- Manipulating inequalities   49
- Solving linear inequalities   49
- Solving quadratic inequalities   50
- Intersection of a line and a parabola   51

Chapter 7 Differentiation   54

- Chords, tangents, normals and gradients   54
- The gradient at any point on the curve $y = x^2$ 55
- Differentiating x^n with respect to x 56
- Differentiating constants and multiples of x 56
- Gradients of tangents and normals   57
- Increasing and decreasing functions   59
- The chain rule   60
- Connected rates of change   62

Chapter 8 Tangents, normals and stationary values   65

- The equations of tangents and normals   65
- Stationary values   67
- Turning points   68

Summary 2   75

Chapter 9 Trigonometry 1   79

- The trigonometric functions   79
- The trigonometric ratios of 30°, 45°, 60°   80
- The sine function   80
- The cosine function   86
- The tangent function   87
- Relationships between $\sin \theta$, $\cos \theta$ and $\tan \theta$ 89

Chapter 10 Trigonometry 2   92

- Trigonometric identities   92
- The Pythagorean identity   92
- Solving equations   94
- Equations involving multiple angles   96
- The inverse trigonometric functions   97

Chapter 11 Sequences and series   100

- Arithmetic progressions   100
- The sum of an arithmetic progression   101

- Geometric progressions 103
- The sum of the first n terms of a geometric progression ... 104
- The sum to infinity of a geometric progression 105
- The binomial theorem 107

Chapter 12 Integration   110

- Differentiation reversed 110
- Using integration to find an area 113
- Definite integration 114
- Finding area by definite integration 115
- Volume of revolution 122

Chapter 13 Vectors   127

- Vectors 127
- Properties of vectors 127
- Position vectors and displacement vectors 128
- The location of a point in three dimensions 129
- Operations on Cartesian vectors 130
- Finding a unit vector 133
- The angle between two vectors 134
- The scalar product 134
- Calculating $\mathbf{a} \cdot \mathbf{b}$ in Cartesian form 135

Summary 3   138

List of formulae   143

Sample examination papers   145

Answers   149

Index   167

Introduction

The *Nelson Mathematics for Cambridge International A Level* series has been written specifically for students of Cambridge's 9709 syllabus by an experienced author team in collaboration with examiners who are very familiar with the syllabus and examinations. This means that, no matter which combination of modules you have chosen, the content of this series matches the content of the syllabus exactly and will give you firm guidelines on which to base your studies.

In this book, the content of the Pure Mathematics 1 module is divided into 13 chapters that give a sensible order for your studies. The chapters begin with a list of objectives that show you what is covered.

The following features help you to understand the concepts of the P1 module and to succeed in your exams.

- The introductions to concepts are accompanied by examples of questions together with their solutions. These show each step of working along with a commentary on the reasoning processes involved.

- There are numerous exercises for you to practise what you have learned and develop your skills.

- There are three Summary exercise sections with more detailed questions covering the content of the preceding chapters. These questions are similar to those found in exam papers and all are from real exam papers.

- Summaries of key information and formulae are at convenient points in the book to help you revise what you have covered in the last few chapters.

- Answers to all questions are provided at the back of the book for you to check your answers to exercises.

- Two sample examination papers have been created in the style of Cambridge's International A Level P1 exams to give you the experience of working through a full examination paper.

1 Quadratic equations

After studying this chapter you should be able to

- carry out the process of completing the square for a quadratic polynomial $ax^2 + bx + c$
- find the discriminant of a quadratic polynomial $ax^2 + bx + c$ and use the discriminant, e.g. to determine the number of real roots of the equation $ax^2 + bx + c = 0$
- solve quadratic equations
- solve by substitution a pair of simultaneous equations of which one is linear and one is quadratic
- recognise and solve equations in x that are quadratic in some function of x, e.g. $x^4 - 5x^2 + 4 = 0$

EXPRESSIONS AND EQUATIONS

An expression is a relationship between terms. For example, $3x - 2$ and $\dfrac{2x - 1}{2 + x}$ are expressions.

An equation is formed when two expressions are equal. For example, $3x - 2 = x - 1$ and $\dfrac{2x - 1}{2 + x} = 5x$ are equations.

An equation has solutions. For example, $x = \frac{1}{2}$ is the solution of the equation $3x - 2 = x - 1$ because, when $x = \frac{1}{2}$ the two expressions are equal.

QUADRATIC EQUATIONS

Any quadratic equation can be written as $ax^2 + bx + c = 0$ where x is a variable and a, b and c are constants and $a \neq 0$.

SOLUTION BY FACTORISING

The left-hand side of the quadratic equation $x^2 - 3x + 2 = 0$ can be factorised,

i.e. $x^2 - 3x + 2 = (x - 2)(x - 1)$

Therefore the given equation becomes

$$(x - 2)(x - 1) = 0 \qquad\qquad [1]$$

When the product of two quantities is zero then one, or both, of those quantities must be zero.

Applying this fact to equation [1] gives

$$x - 2 = 0 \qquad \text{or} \qquad x - 1 = 0$$

i.e. $x = 2$ or $x = 1$

This is the solution of the given equation.

The values 2 and 1 are called the *roots* of that equation.

This method can be used for any quadratic equation in which the quadratic expression factorises.

Example 1a

Find the roots of the equation $x^2 + 6x - 7 = 0$

$$x^2 + 6x - 7 = 0$$
$$\Rightarrow \qquad (x - 1)(x + 7) = 0$$
$$\therefore \qquad x - 1 = 0 \text{ or } x + 7 = 0$$
$$\therefore \qquad x = 1 \text{ or } x = -7$$

The roots of the equation are 1 and -7.

Exercise 1a

Solve the equations.

1 $x^2 + 5x + 6 = 0$

2 $x^2 + x - 6 = 0$

3 $x^2 - x - 6 = 0$

4 $x^2 + 6x + 8 = 0$

5 $x^2 - 4x + 3 = 0$

6 $x^2 + 2x - 3 = 0$

7 $2x^2 + 3x + 1 = 0$

8 $4x^2 - 9x + 2 = 0$

9 $x^2 + 4x - 5 = 0$

10 $x^2 + x - 72 = 0$

Find the roots of the equations.

11 $x^2 - 2x - 3 = 0$

12 $x^2 + 5x + 4 = 0$

13 $x^2 - 6x + 5 = 0$

14 $x^2 + 3x - 10 = 0$

15 $x^2 - 5x - 14 = 0$

16 $x^2 - 9x + 14 = 0$

Rearranging the equation

The terms in a quadratic equation are not always given in the order $ax^2 + bx + c = 0$. When they are given in a different order they should be rearranged into the standard form.

For example $x^2 - x = 4$ becomes $x^2 - x - 4 = 0$

$3x^2 - 1 = 2x$ becomes $3x^2 - 2x - 1 = 0$

$x(x - 1) = 2$ becomes $x^2 - x = 2 \Rightarrow x^2 - x - 2 = 0$

Also collect the terms on the side where the x^2 term is positive, for example

$2 - x^2 = 5x$ becomes $0 = x^2 + 5x - 2$

i.e. $x^2 + 5x - 2 = 0$

Example 1b

Solve the equation $4x - x^2 = 3$

$$4x - x^2 = 3$$
$$\Rightarrow \qquad 0 = x^2 - 4x + 3$$
$$\Rightarrow \qquad x^2 - 4x + 3 = 0 \Rightarrow (x - 3)(x - 1) = 0$$
$$\Rightarrow \qquad x - 3 = 0 \quad \text{or} \quad x - 1 = 0$$
$$\Rightarrow \qquad x = 3 \quad \text{or} \quad x = 1$$

Losing a solution

Quadratic equations sometimes have a common factor containing the unknown quantity. It is very tempting in such cases to divide by the common factor, but doing this results in the loss of part of the solution, as the following example shows.

Correct solution

$$x^2 - 5x = 0$$
$$x(x - 5) = 0$$
$$\therefore \qquad x = 0 \text{ or } x - 5 = 0$$
$$\Rightarrow \qquad x = 0 \text{ or } 5$$

Faulty solution

$$x^2 - 5x = 0$$
$$x - 5 = 0 \quad \text{(Dividing by } x\text{)}$$
$$\therefore \qquad x = 5$$

The solution $x = 0$ has been lost.

Dividing an equation by a numerical common factor is correct and sensible; dividing by a common factor containing the unknown quantity results in the loss of a solution.

Exercise 1b

Solve each equation, making sure that you give all the roots.

1 $x^2 + 10 - 7x = 0$

2 $15 - x^2 - 2x = 0$

3 $x^2 - 3x = 4$

4 $12 - 7x + x^2 = 0$

5 $2x - 1 + 3x^2 = 0$

6 $x(x + 7) + 6 = 0$

7 $2x^2 - 4x = 0$

8 $x(4x + 5) = -1$

9 $2 - x = 3x^2$

10 $6x^2 + 3x = 0$

11 $x^2 + 6x = 0$

12 $x^2 = 10x$

13 $x(4x + 1) = 3x$

14 $20 + x(1 - x) = 0$

15 $x(3x - 2) = 8$

16 $x^2 - x(2x - 1) + 2 = 0$

17 $x(x + 1) = 2x$

18 $4 + x^2 = 2(x + 2)$

19 $x(x - 2) = 3$

20 $1 - x^2 = x(1 + x)$

COMPLETING THE SQUARE

We can make a perfect square by adding a constant to the x^2 and x term of a quadratic expression. This is called completing the square.

For example, $x^2 - 2x$: adding 1 gives $x^2 - 2x + 1$

and $x^2 - 2x + 1 = (x - 1)^2$ which is a perfect square.

Adding the number 1 was not a guess, it was found by using the fact that

$$x^2 + 2ax + \boxed{a^2} = (x + a)^2$$

Hence the number to be added is always (half the coefficient of x)^2.

$\therefore$ $x^2 + 6x$ needs 3^2 to be added to make a perfect square,

i.e. $x^2 + 6x + 9 = (x + 3)^2$

To complete the square when the coefficient of x^2 is not 1, first take out the coefficient of x^2 as a factor,

e.g. $2x^2 + x = 2(x^2 + \frac{1}{2}x)$

Now add $(\frac{1}{2} \times \frac{1}{2})^2$ inside the bracket, giving

$$2(x^2 + \tfrac{1}{2}x + \tfrac{1}{16}) = 2(x + \tfrac{1}{4})^2$$

When the coefficient of x^2 is negative, take -1 out as a factor,

e.g. $\qquad -x^2 + 4x = -(x^2 - 4x)$

Then $\qquad -(x^2 - 4x + 4) = -(x - 2)^2, \quad \therefore -x^2 + 4x - 4 \equiv -(x - 2)^2$

Completing the square can be used to solve a quadratic equation.

Examples 1c

1 Solve the equation $x^2 - 4x - 2 = 0$

$x^2 - 4x - 2 = 0 \Rightarrow x^2 - 4x = 2$

No factors can be found, so isolate the two terms with x in.

Add $\left\{\frac{1}{2} \times (-4)\right\}^2$ to *both* sides, i.e. $x^2 - 4x + 4 = 2 + 4$

$\Rightarrow \quad (x - 2)^2 = 6$, i.e. $x - 2 = \pm\sqrt{6} = \pm 2.4494\ldots$

$\therefore \quad x = -0.449$ or 4.45 (3 s.f.)

2 Find the roots of the equation $2x^2 - 3x - 3 = 0$

$2\left(x^2 - \tfrac{3}{2}x\right) = 3 \Rightarrow x^2 - \tfrac{3}{2}x = \tfrac{3}{2}$

$x^2 - \tfrac{3}{2}x + \tfrac{9}{16} = \tfrac{3}{2} + \tfrac{9}{16} \Rightarrow \left(x - \tfrac{3}{4}\right)^2 = \tfrac{33}{16}$

$x - 0.75 = \pm 1.4361\ldots$

the roots are -0.686 and 2.19 (3 s.f.)

Exercise 1c

Add a number to each expression so that the result is a perfect square.

1 $x^2 - 4x$

2 $x^2 + 2x$

3 $x^2 - 6x$

4 $x^2 + 10x$

5 $2x^2 - 4x$

6 $x^2 + 5x$

7 $3x^2 - 48x$

8 $x^2 + 18x$

9 $2x^2 - 40x$

10 $x^2 + x$

11 $3x^2 - 2x$

12 $2x^2 + 3x$

Solve the equations by completing the square.

13 $x^2 + 8x = 1$

14 $x^2 - 2x - 2 = 0$

15 $x^2 + x - 1 = 0$

16 $2x^2 + 2x = 1$

17 $x^2 + 3x + 1 = 0$

18 $2x^2 - x - 2 = 0$

19 $x^2 + 4x = 2$

20 $3x^2 + x - 1 = 0$

21 $2x^2 + 4x = 7$

22 $x^2 - x = 3$

23 $4x^2 + x - 1 = 0$

24 $2x^2 - 3x - 4 = 0$

THE FORMULA FOR SOLVING A QUADRATIC EQUATION

Completing the square to solve the general quadratic equation gives a formula that can be used to solve any quadratic equation.

Using completing the square for $ax^2 + bx + c = 0$ gives

$$ax^2 + bx = -c$$

i.e.
$$a\left(x^2 + \frac{b}{a}x\right) = -c$$

$\Rightarrow$
$$x^2 + \frac{b}{a}x = -\frac{c}{a}$$

$\therefore$
$$x^2 + \frac{b}{a}x + \left(\frac{b}{2a}\right)^2 = \left(\frac{b}{2a}\right)^2 - \frac{c}{a}$$

$\therefore$
$$\left(x + \frac{b}{2a}\right)^2 = \frac{b^2}{4a^2} - \frac{c}{a} = \frac{b^2 - 4ac}{4a^2}$$

$\Rightarrow$
$$x + \frac{b}{2a} = \pm\sqrt{\frac{b^2 - 4ac}{4a^2}}$$

$\Rightarrow$
$$x = -\frac{b}{2a} \pm \frac{\sqrt{b^2 - 4ac}}{2a}$$

i.e.
$$x = \frac{-b \pm \sqrt{b^2 - 4ac}}{2a}$$

Example 1d

Use the formula to find the roots of the equation $2x^2 - 7x - 1 = 0$

$2x^2 - 7x - 1 = 0$

Comparing with $ax^2 + bx + c = 0$ gives $a = 2, b = -7, c = -1$

$$x = \frac{-b \pm \sqrt{b^2 - 4ac}}{2a} = \frac{7 \pm \sqrt{49 - 4(2)(-1)}}{4}$$

Therefore, $x = \dfrac{7 \pm \sqrt{57}}{4}$

The roots are 3.64 and -0.137 (3 s.f.).

Exercise 1d

Use the formula to solve the equations.

1 $x^2 + 4x + 2 = 0$

2 $2x^2 - x - 2 = 0$

3 $x^2 + 5x + 1 = 0$

4 $2x^2 - x - 4 = 0$

5 $x^2 + 1 = 4x$

6 $2x^2 - x = 5$

7 $1 + x - 3x^2 = 0$

8 $3x^2 = 1 - x$

Find the roots of the equations.

9 $5x^2 + 9x + 2 = 0$

10 $2x^2 - 7x + 4 = 0$

11 $4x^2 - 7x - 1 = 0$

12 $3x = 5 - 4x^2$

13 $4x^2 + 3x = 5$

14 $1 = 5x - 5x^2$

15 $8x - x^2 = 1$

16 $x^2 - 3x = 1$

EQUATIONS THAT ARE QUADRATIC IN A FUNCTION OF x

An equation that is quadratic in a function of x can be reduced to a quadratic equation in x by replacing the function of x by y.

For example, replacing x^2 by y in the equation $x^4 + x^2 - 12 = 0$ gives $y^2 + y - 12 = 0$

Then	$y^2 + y - 12 = 0 \Rightarrow (y-3)(y+4) = 0$
So	$y = 3$ or $y = -4$
$\therefore$	$x^2 = 3$ or $x^2 = -4$

There are no real values of x for which $x^2 = -4$, so $x^2 = 3 \Rightarrow x = \pm 1.73$

Example 1e

Solve the equation $\dfrac{1}{x^2} - \dfrac{13}{x} + 36 = 0$

Provided that $x \neq 0$, replacing $\dfrac{1}{x}$ by y gives $y^2 - 13y + 36 = 0$

This is needed because $\dfrac{1}{x}$ is meaningless if $x = 0$

$y^2 - 13y + 36 = 0 \Rightarrow (y - 9)(y - 4) = 0$

$\therefore y = 9$ or $y = 4$ so $\dfrac{1}{x} = 9$ or $\dfrac{1}{x} = 4$

$\Rightarrow x = \dfrac{1}{9}$ or $x = \dfrac{1}{4}$

Exercise 1e

Solve the equations.

1   $x^2 + x^4 = 20$

2   $x^2(x^2 + 1) = 12$

3   $x^4 = 4x^2 - 4$

4   $(4x^2 - 5)(x^2 - 2) = 10$

5   $16x^4 = 81$

6   $8x^4 - 4x^2 = 0$

7   $\dfrac{1}{x^2} - \dfrac{1}{x} = 30$

8   $9 = \dfrac{6}{x^2} - \dfrac{1}{x^4}$

9   $\dfrac{1}{x^4} - \dfrac{7}{x^2} + 10 = 0$

10   $\dfrac{3}{x^4} - \dfrac{1}{x^2} = 2$

11   $(x^2 - 1)^2 - 5(x^2 - 1) + 4 = 0$

12   $(x^2 - 4)^2 - 9 = 0$

13   $x - 3x^{\frac{1}{2}} + 2 = 0$

14   $2x + 5\sqrt{x} - 12 = 0$

15   $\dfrac{1}{(x^2 - 4)^2} - \dfrac{1}{x^2 - 4} - 20 = 0$

16   Explain why the equation $3x^4 + 5x^2 + 2 = 0$ has no real roots.

PROPERTIES OF THE ROOTS OF A QUADRATIC EQUATION

A number of facts can be found from the formula used for solving a quadratic equation.

$$x = -\frac{b}{2a} \pm \frac{\sqrt{b^2 - 4ac}}{2a}$$

The sum of the roots

The separate roots are $-\dfrac{b}{2a} + \dfrac{\sqrt{b^2 - 4ac}}{2a}$ and $-\dfrac{b}{2a} - \dfrac{\sqrt{b^2 - 4ac}}{2a}$

When the roots are added, the terms containing the square root disappear, giving

$$\textbf{sum of roots} = -\frac{b}{a}$$

This fact gives a useful check on the accuracy of roots that have been calculated.

The nature of the roots

In the formula there are two terms. The first of these, $-\dfrac{b}{2a}$, can always be found for any values of a and b.

For the second term, i.e. $\dfrac{\sqrt{b^2 - 4ac}}{2a}$, there are three different cases to consider.

1 If $b^2 - 4ac$ is positive, its square root can be found and it is a *real* number.

   The two square roots, i.e. $\pm \sqrt{b^2 - 4ac}$ have different (or distinct) values giving two different real values of x. So the equation has *two different real roots*.

2 If $b^2 - 4ac$ is zero then its square root also is zero and $x = -\dfrac{b}{2a} - \dfrac{\sqrt{b^2 - 4ac}}{2a}$ gives

   $$x = -\frac{b}{2a} + 0 \text{ and } x = -\frac{b}{2a} - 0$$

   i.e. there is just one value of x that satisfies the equation.

   For an example $x^2 - 2x + 1 = 0$

   From the formula we get $x = -\dfrac{(-2)}{2} \pm 0$,   i.e. $x = 1$ or 1

   By factorising we can see that there are two equal roots,

   i.e. $(x - 1)(x - 1) = 0 \Rightarrow x = 1$ or $x = 1$

   This type of equation has a *repeated root*.

3 If $b^2 - 4ac$ is negative we cannot find its square root because there is no real number whose square is negative. In this case the equation has *no real roots*.

   Therefore the roots of a quadratic equation can be

   either   real and different

   or   real and equal

   or   not real

   and it is the value of $b^2 - 4ac$ that determines the nature of the roots.

$$\boldsymbol{b^2 - 4ac} \textbf{ is called the discriminant.}$$

Condition	Nature of Roots
$b^2 - 4ac > 0$	Real and different
$b^2 - 4ac = 0$	Real and equal
$b^2 - 4ac < 0$	Not real

Sometimes it matters only that the roots are real, in which case the first two conditions can be combined to give

$$\textbf{if } \boldsymbol{b^2 - 4ac \geqslant 0} \textbf{ the roots are real.}$$

Examples 1f

1 Find the nature of the roots of the equation $x^2 - 6x + 1 = 0$

$x^2 - 6x + 1 = 0$

$a = 1, b = -6, c = 1$

$b^2 - 4ac = (-6)^2 - 4(1)(1) = 32$

$b^2 - 4ac > 0$ so the roots are real and different.

2 The roots of the equation $2x^2 - px + 8 = 0$ are equal. Find the value of p.

$2x^2 - px + 8 = 0$

$a = 2, b = -p, c = 8$

The roots are equal so $b^2 - 4ac = 0$

i.e. $\qquad (-p)^2 - 4(2)(8) = 0$

$\Rightarrow \qquad\qquad p^2 - 64 = 0 \quad \Rightarrow \quad p^2 = 64 \quad \therefore \quad p = \pm 8$

3 Prove that the equation $(k - 2)x^2 + 2x - k = 0$ has real roots whatever the value of k.

$(k - 2)x^2 + 2x - k = 0$

$a = k - 2, b = 2, c = -k$

$b^2 - 4ac = 4 - 4(k - 2)(-k) = 4 + 4k^2 - 8k$

$\qquad\qquad = 4k^2 - 8k + 4 = 4(k^2 \quad 2k + 1) - 4(k - 1)^2$

$(k - 1)^2$ cannot be negative whatever the value of k, so $b^2 - 4ac$ cannot be negative. Therefore the roots are always real.

Exercise 1f

Without solving the equation, write down the sum of its roots.

1 $x^2 - 4x - 7 = 0$

2 $3x^2 + 5x + 1 = 0$

3 $2 + x - x^2 = 0$

4 $3x^2 - 4x - 2 = 0$

5 $x^2 + 3x + 1 = 0$

6 $7 + 2x - 5x^2 = 0$

In questions **7** to **16**, without solving the equation, determine the nature of its roots.

7 $x^2 - 6x + 4 = 0$

8 $3x^2 + 4x + 2 = 0$

9 $2x^2 - 5x + 3 = 0$

10 $x^2 - 6x + 9 = 0$

11 $4x^2 - 12x - 9 = 0$

12 $4x^2 + 12x + 9 = 0$

13 $x^2 + 4x - 8 = 0$

14 $x^2 + ax + a^2 = 0$

15 $x^2 - ax - a^2 = 0$

16 $x^2 + 2ax + a^2 = 0$

17 The roots of $3x^2 + kx + 12 = 0$ are equal. Find k.

18 $x^2 - 3x + a = 0$ has equal roots. Find a.

19 The roots of $x^2 + px + (p - 1) = 0$ are equal. Find p.

20 Prove that the roots of the equation $kx^2 + (2k + 4)x + 8 = 0$ are real for all values of k.

21 Show that the equation $ax^2 + (a + b)x + b = 0$ has real roots for all values of a and b.

22 Find the relationship between p and q if the roots of the equation $px^2 + qx + 1 = 0$ are equal.

SOLUTION OF ONE LINEAR AND ONE QUADRATIC EQUATION

We can solve a pair of simultaneous equations where one is linear and the other is quadratic by substitution. We use the linear equation to express one unknown in terms of the other and substitute this in the quadratic equation.

Example 1g

Solve the equations $x - y = 2$
$$2x^2 - 3y^2 = 15$$

$$x - y = 2 \qquad [1]$$
$$2x^2 - 3y^2 = 15 \qquad [2]$$

Equation [1] is linear so use it for the substitution.

$$[1] \Rightarrow x = y + 2$$

Substituting $y + 2$ for x in [2] gives

$$2(y + 2)^2 - 3y^2 = 15$$
$$\Rightarrow \qquad 2(y^2 + 4y + 4) - 3y^2 = 15$$
$$\Rightarrow \qquad 2y^2 + 8y + 8 - 3y^2 = 15$$

Collecting terms on the side where y^2 is positive gives

$$0 = y^2 - 8y + 7$$
$$\Rightarrow \qquad 0 = (y - 7)(y - 1), \ y = 7 \text{ or } 1$$

We use $x = y + 2$ to find corresponding values of x.

y	7	1
x	9	3

$\therefore$ either $x = 9$ and $y = 7$ or $x = 3$ and $y = 1$

The values of x and y must be given in *corresponding pairs*. It is wrong to write the answer as $y = 7$ or 1 and $x = 9$ or 3 because $y = 7$ with $x = 3$ and $y = 1$ with $x = 9$ are *not* solutions.

Exercise 1g

Solve the following pairs of equations.

1 $x^2 + y^2 = 5$
    $y - x = 1$

2 $y^2 - x^2 = 8$
    $x + y = 2$

3 $3x^2 - y^2 = 3$
    $2x - y = 1$

4      $y = 4x^2$
    $y + 2x = 2$

5 $y^2 + xy = 3$
  $2x + y = 1$

6 $x^2 - xy = 14$
  $y = 3 - x$

7 $xy = 2$
  $x + y - 3 = 0$

8 $2x - y = 2$
  $x^2 - y = 5$

9 $y - x = 4$
  $y^2 - 5x^2 = 20$

10 $x + y^2 = 10$
  $x - 2y = 2$

11 $4x + y = 1$
  $4x^2 + y = 0$

12 $3xy - x = 0$
  $x + 3y = 2$

13 $x^2 + 4y^2 = 2$
  $2y + x + 2 = 0$

14 $x + 3y = 0$
  $2x + 3xy = 1$

15 $3x - 4y = 1$
  $6xy = 1$

16 $x^2 + 4y^2 = 2$
  $x + 2y = 2$

17 $xy = 9$
  $x - 2y = 3$

18 $4x + y = 2$
  $4x + y^2 = 8$

19 $1 + 3xy = 0$
  $x + 6y = 1$

20 $x^2 - xy = 0$
  $x + y = 1$

21 $xy + y^2 = 2$
  $2x + y = 3$

22 $xy + x = -3$
  $2x + 5y = 8$

Mixed exercise 1

Solve the equations.

1 $x^2 - 5x - 6 = 0$

2 $x^2 - 6x - 5 = 0$

3 $2x^2 + 3x = 1$

4 $5 - 3x^2 = 4x$

5 $x(2 - x) = 1$

6 $4x^2 - 3 = 11x$

7 $(x - 1)(x + 2) = 1$

8 $x^2 + 4x + 4 = 16$

9 $x^2 + 2x = 2$

10 $2(x^2 + 2) = x(x - 4)$

In questions **11** to **16**, solve the equations giving all possible solutions.

11 $x(x - 2) = 0$

12 $x(x - 5) = 2(x + 5)$

13 $x^2(x^2 + 3) = 0$

14 $\dfrac{1}{x^4} - \dfrac{2}{x^2} + 1 = 0$

15 $x + 3\sqrt{x} = 6$

16 $(x^2 - 1)^2 - 6(x^2 - 1) + 8 = 0$

In questions **17** and **18** solve the pair of equations.

17 $2x^2 - y^2 = 7$
  $x + y = 9$

18 $2x = y - 1$
  $x^2 - 3y + 11 = 0$

19 For each equation, first find the value of $-\dfrac{b}{a}$, then use any method to find the roots of the equation and finally find the sum of the roots and check that it is equal to $-\dfrac{b}{a}$.

(a) $x^2 - 6x + 8 = 0$

(b) $4x^2 + 5x = 3$

20 Determine the nature of the roots of the equations.

(a) $x^2 + 3x + 7 = 0$

(b) $3x^2 - x - 5 = 0$

(c) $ax^2 + 2ax + a = 0$

(d) $2 + 9x - x^2 = 0$

21 For what values of p does the equation $px^2 + 4x + (p - 3) = 0$ have equal roots?

22 Show that the equation $2x^2 + 2(p + 1)x + p = 0$ always has real roots.

23 The equation $x^2 + kx + k = 1$ has equal roots. Find k.

2 Coordinate geometry 1

After studying this chapter you should be able to

- find the length, gradient and mid-point of a line segment, given the coordinates of the end-points
- understand and use the relationships between the gradients of parallel and perpendicular lines.

CARTESIAN COORDINATES

Cartesian coordinates are used to give the position of a point in a plane.

There are two perpendicular lines called the *x*-axis, Ox, and the *y*-axis, Oy. The axes cross at a point O, called the origin.

The position of the point A is given as the ordered pair $(3, 2)$ where 3 is the *x*-coordinate and 2 is the *y*-coordinate.

The *x*-coordinate gives the distance of a point from O parallel to the *x*-axis and the *y*-coordinate gives the distance of a point from O parallel to the *y*-axis.

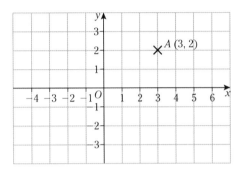

THE LENGTH OF A LINE JOINING TWO POINTS

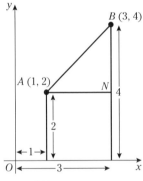

The length of the line joining the points $A(1, 2)$ and $B(3, 4)$ can be found by using Pythagoras' theorem.

$$\Rightarrow \quad AB^2 = AN^2 + BN^2$$
$$= (3 - 1)^2 + (4 - 2)^2$$
$$= 8$$

Therefore $AB = \sqrt{8} = 2.83$ (3 s.f.)

The length of the line joining any two points can also be found by using Pythagoras' theorem.

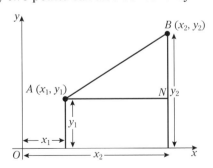

From the diagram, $AB^2 = AN^2 + BN^2$

$$= (x_2 - x_1)^2 + (y_2 - y_1)^2$$
$$\Rightarrow \quad AB = \sqrt{(x_2 - x_1)^2 + (y_2 - y_1)^2}$$

the length of the line joining $A(x_1, y_1)$ to $B(x_1, y_1)$ is given by

$$AB = \sqrt{(x_2 - x_1)^2 + (y_2 - y_1)^2}$$

Examples 2a

1 Find the length of the line joining $A(-2, 2)$ to $B(3, -1)$.

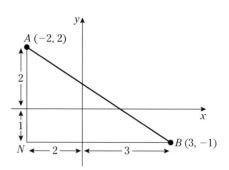

$$AB = \sqrt{(x_2 - x_1)^2 + (y_2 - y_1)^2}$$
$$= \sqrt{(3 - \{-2\})^2 + (-1 - 2)^2}$$
$$= \sqrt{5^2 + (-3)^2}$$
$$= \sqrt{34} = 5.83 \ (3\,\text{s.f.})$$

From the diagram, $BN = 3 + 2 = 5$ and $AN = 2 + 1 = 3$

$\Rightarrow \qquad AB^2 = 5^2 + 3^2 = 34 \Rightarrow AB = \sqrt{34}$

This confirms that the formula used above also works when some of the coordinates are negative.

THE MIDPOINT OF A LINE JOINING TWO POINTS

M is the midpoint of the line joining $A(1, 1)$ and $B(5, 3)$.

The diagram shows that triangles AMN and MBT are congruent.

Therefore $AN = MT$, showing that N is the midpoint of AF.

So S is the midpoint of CD.

Therefore the x-coordinate of M is given by OS, where

$$OS = OC + \tfrac{1}{2}CD = 1 + \tfrac{1}{2}(5 - 1) = 3$$

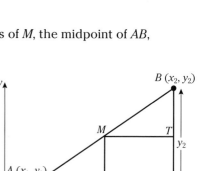

Also, T is the midpoint of BF, so the y-coordinate of M is given by $SM\ (= DT)$, where

$$DT = DF + \tfrac{1}{2}FB = 1 + \tfrac{1}{2}(3 - 1) = 2$$

Therefore M is the point $(3, 2)$.

When $A(x_1, y_1)$ and $B(x_2, y_2)$ are any two points, then the coordinates of M, the midpoint of AB, can be found in the same way.

At M, $\qquad x = OS = OR + \tfrac{1}{2}RU$

$$\qquad\qquad = x_1 + \tfrac{1}{2}(x_2 - x_1) = \tfrac{1}{2}(x_1 + x_2)$$

and $\qquad y = SM = UT = UV + \tfrac{1}{2}BV$

$$\qquad\qquad = y_1 + \tfrac{1}{2}(y_2 - y_1) = \tfrac{1}{2}(y_1 + y_2)$$

The coordinates of the midpoint of the line joining $A(x_1, y_1)$ and $B(x_2, y_2)$ are $\left[\tfrac{1}{2}(x_1 + x_2), \tfrac{1}{2}(y_1 + y_2)\right]$

These coordinates are easy to remember as the average of the coordinates of A and B.

Examples 2a cont.

2 Find the coordinates of the midpoint of the line joining $A(-3, -2)$ and $B(1, 3)$.

The coordinates of M are $\left[\frac{1}{2}(x_1 + x_2), \frac{1}{2}(y_1 + y_2)\right]$

$$= \left[\frac{1}{2}(-3 + 1), \frac{1}{2}(-2 + 3)\right] = \left(-1, \frac{1}{2}\right)$$

Alternatively, from the diagram, M is half-way from A to B horizontally and vertically,

i.e. at M $x = -3 + \frac{1}{2}(4) = -1$ and $y = -2 + \frac{1}{2}(5) = \frac{1}{2}$

This confirms that the formula also works when some of the coordinates are negative.

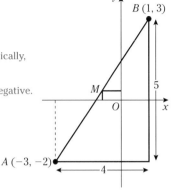

Exercise 2a

1 Find the length of the line joining
   (a) $A(1, 2)$ and $B(4, 6)$
   (b) $C(3, 1)$ and $D(2, 0)$
   (c) $J(4, 2)$ and $K(2, 5)$.

2 Find the coordinates of the midpoints of the lines joining the points in question 1.

3 Find (i) the length, (ii) the coordinates of the midpoint, of the line joining
   (a) $A(-1, -4)$, $B(2, 6)$
   (b) $S(0, 0)$, $T(-1, -2)$
   (c) $E(-1, -4)$, $F(-3, -2)$.

4 Find the distance from the origin to the point $(7, 4)$.

5 Find the length of the line joining the point $(-3, 2)$ to the origin.

6 Find the coordinates of the midpoint of the line from the point $(4, -8)$ to the origin.

7 Show, by using Pythagoras' theorem, that the lines joining $A(1, 6)$, $B(-1, 4)$ and $C(2, 1)$ form a right-angled triangle.

8 A, B and C are points $(7, 3)$, $(-4, 1)$ and $(-3, -2)$ respectively.
   (a) Show that $\triangle ABC$ is isosceles.
   (b) Find the midpoint of BC.
   (c) Find the area of $\triangle ABC$.

9 The vertices of a triangle are $A(0, 2)$, $B(1, 5)$ and $C(-1, 4)$. Find
   (a) the perimeter of the triangle
   (b) the coordinates of D such that AD is a median of $\triangle ABC$

   The median of a triangle is the line from a vertex to the midpoint of the opposite side.

   (c) the length of AD.

10 Show that the lines OA and OB are perpendicular where A and B are the points $(4, 3)$ and $(3, -4)$ respectively.

11 M is the midpoint of the line joining A to B. The coordinates of A and M are $(5, 7)$ and $(0, 2)$ respectively. Find the coordinates of B.

GRADIENT

The gradient of a straight line is a measure of its inclination with respect to the x-axis.
Gradient is defined as

the increase in y divided by the increase in x between one point and another point on the line.

For the line passing through $A(1, 2)$ and $B(4, 3)$,

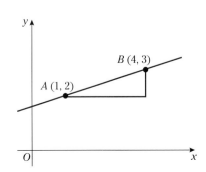

from A to B,   the increase in y is 1,
                      the increase in x is 3.

Therefore the gradient of AB is $\frac{1}{3}$.

The gradient of a line may be found from *any* two points on the line.

For example, for the line through the points $A(2, 3)$ and $B(6, 1)$,

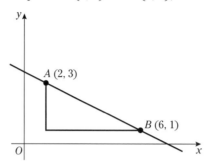

moving from A to B $\dfrac{\text{increase in } y}{\text{increase in } x} = \dfrac{-2}{4} = -\dfrac{1}{2}$

Alternatively, moving from B to A $\dfrac{\text{increase in } y}{\text{increase in } x} = \dfrac{2}{-4} = -\dfrac{1}{2}$

Therefore it does not matter in which order the two points are used, provided that their coordinates are used in the *same* order when calculating the increases in x and in y.

These two examples show that the gradient of a line may be positive or negative.

A positive gradient shows that the line makes an acute angle with the positive direction of the x-axis.

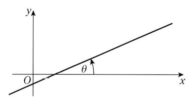

A negative gradient shows that the line makes an obtuse angle with the positive direction of the x-axis.

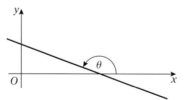

The gradient of the line passing through any two points $A(x_1, y_1)$ and $B(x_2, y_2)$ is

$$\frac{\textbf{the increase in } \textbf{\textit{y}}}{\textbf{the increase in } \textbf{\textit{x}}} = \frac{\textbf{\textit{y}}_2 - \textbf{\textit{y}}_1}{\textbf{\textit{x}}_2 - \textbf{\textit{x}}_1}$$

As the gradient of a straight line is the increase in y divided by the increase in x from one point on the line to another,

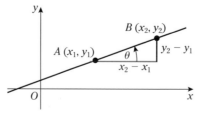

gradient measures the increase in y per unit increase in x,
i.e. the rate of increase of y with respect to x.

PARALLEL LINES

If l_1 and l_2 are parallel lines, they are equally inclined to the positive direction of the x-axis,

i.e.   **parallel lines have equal gradients.**

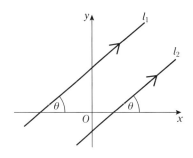

PERPENDICULAR LINES

The perpendicular lines AB and CD have gradients m_1 and m_2 respectively.

AB makes an angle θ with the x-axis so CD makes an angle θ with the y-axis.

Therefore triangles PQR and PST are similar.

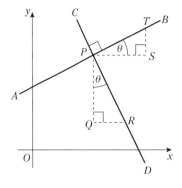

The gradient of AB is $\dfrac{ST}{PS} = m_1$

and the gradient of CD is $\dfrac{-PQ}{QR} = m_2$, i.e. $\dfrac{PQ}{QR} = -m_2$

But $\dfrac{ST}{PS} = \dfrac{QR}{PQ}$ (triangles PQR and PST are similar)

therefore $m_1 = -\dfrac{1}{m_2}$ or $m_1 m_2 = -1$

The product of the gradients of perpendicular lines is -1,

i.e. if one line has gradient m, any perpendicular line has gradient $-\dfrac{1}{m}$.

Example 2b

Determine, by comparing gradients, whether the following three points are collinear (i.e. lie on the same straight line).

$A(\frac{2}{3}, 1), B(1, \frac{1}{2}), C(2, -1)$

The gradient of AB is $\dfrac{1 - \frac{1}{2}}{\frac{2}{3} - 1} = -\dfrac{3}{2}$

The gradient of BC is $\dfrac{-1 - \frac{1}{2}}{2 - 1} = -\dfrac{3}{2}$

As the gradients of AB and BC are the same, A, B and C are collinear.

The diagram, although not strictly necessary, gives a check that the answer is reasonable.

Exercise 2b

1 Find the gradient of the line through the pairs of points.

   (a) $(0, 0), (1, 3)$

   (b) $(1, 4), (3, 7)$

   (c) $(5, 4), (2, 3)$

   (d) $(-1, 4), (3, 7)$

   (e) $(-1, -3), (-2, 1)$

   (f) $(-1, -6), (0, 0)$

   (g) $(-2, 5), (1, -2)$

   (h) $(3, -2), (-1, 4)$

   (i) $(h, k), (0, 0)$

2 Determine whether the given points are collinear.

   (a) $(0, -1), (1, 1), (2, 3)$

   (b) $(0, 2), (2, 5), (3, 7)$

   (c) $(-1, 4), (2, 1), (-2, 5)$

   (d) $(0, -3), (1, -4), \left(-\frac{1}{2}, -\frac{5}{2}\right)$

3 Determine whether AB and CD are parallel, perpendicular or neither.

   (a) $A(0, -1), B(1, 1), C(1, 5), D(-1, 1)$

   (b) $A(1, 1), B(3, 2), C(-1, 1), D(0, -1)$

   (c) $A(3, 3), B(-3, 1), C(-1, -1), D(1, -7)$

   (d) $A(2, -5), B(0, 1), C(-2, 2), D(3, -7)$

   (e) $A(2, 6), B(-1, -9), C(2, 11), D(0, 1)$

MIXED PROBLEMS

Example 2c

The vertices of a triangle are the points $A(2, 4)$, $B(1, -2)$ and $C(-2, 3)$ respectively. The point $H(a, b)$ lies on the altitude through A. Find a relationship between a and b.

An altitude of a triangle is the line from a vertex to the opposite side and perpendicular to the opposite side.

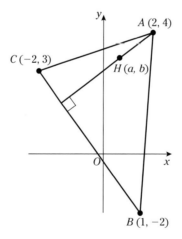

H is on the altitude through A, so AH is perpendicular to BC.

∴ the gradient of AH is $\dfrac{4 - b}{2 - a}$

and the gradient of BC is $\dfrac{3 - (-2)}{-2 - 1} = -\dfrac{5}{3}$

Therefore $\left(\dfrac{4 - b}{2 - a}\right)\left(-\dfrac{5}{3}\right) = -1$ The product of the gradients of perpendicular lines is -1.

⇒ $\dfrac{-20 + 5b}{6 - 3a} = -1$

⇒ $5b = 3a + 14$

Exercise 2c

1

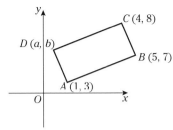

$A(1, 3)$, $B(5, 7)$, $C(4, 8)$, $D(a, b)$ form a rectangle $ABCD$. Find a and b.

2

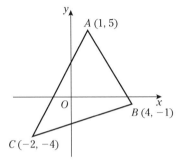

The triangle ABC has its vertices at the points $A(1, 5)$, $B(4, -1)$ and $C(-2, -4)$.

(a)   Show that $\triangle ABC$ is right-angled.

(b)   Find the area of $\triangle ABC$.

3

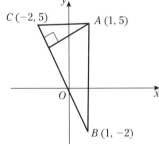

Show that the point $(-\frac{32}{3}, 0)$ is on the perpendicular height through A of the triangle whose vertices are $A(1, 5)$, $B(1, -2)$ and $C(-2, 5)$.

4   Show that the triangle whose vertices are $(1, 1)$, $(3, 2)$, $(2, -1)$ is isosceles.

5   Find, in terms of a and b, the length of the line joining (a, b) and $(2a, 3b)$.

6

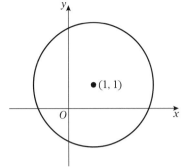

The point $(1, 1)$ is the centre of a circle whose radius is 2. Show that the point $(1, 3)$ is on the circumference of this circle.

7   A circle, radius 2 and centre the origin, cuts the x-axis at A and B and cuts the positive y-axis at C. Prove that $\angle ACB = 90°$.

8   Find in terms of p and q, the coordinates of the midpoint of the line joining $C(p, q)$ and $D(q, p)$. Hence show that the origin is on the perpendicular bisector of the line CD.

9   The point (a, b) is on the circumference of the circle of radius 3 whose centre is at the point $(2, 1)$. Find a relationship between a and b.

10

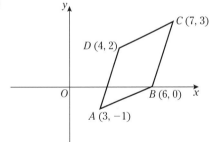

$ABCD$ is a quadrilateral where A, B, C and D are the points $(3, -1)$, $(6, 0)$, $(7, 3)$ and $(4, 2)$. Prove that the diagonals bisect each other at right angles and hence find the area of $ABCD$.

11   The vertices of a triangle are at the points $A(a, 0)$, $B(0, b)$ and $C(c, d)$ and $\angle B = 90°$. Find a relationship between a, b, c and d.

12   A point $P(a, b)$ is equidistant from the y-axis and the point $(4, 0)$. Find a relationship between a and b.

3 Coordinate geometry 2

After studying this chapter you should be able to

- find the equation of a straight line given sufficient information (e.g. the coordinates of two points on it, or one point on it and its gradient)
- interpret and use linear equations, particularly the forms $y = mx + c$ and $y - y_1 = m(x - x_1)$
- understand the relationship between a graph and its associated algebraic equation.

THE EQUATION OF A STRAIGHT LINE

A straight line can be defined in many ways.

- **A line passes through (0, 3) and is parallel to the x-axis.**

  The point $P(x, y)$ is on this line if and only if $y = 3$. Any point whose y-coordinate is not 3 is not on this line.

  $y = 3$ is called the equation of the line.

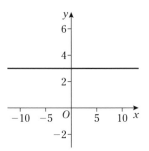

- **A line passes through $(-2, 0)$ and is parallel to the y-axis.**

  The point $P(x, y)$ is on this line if and only if $x = -2$. Any point whose x-coordinate is not -2 is not on this line.

  $x = -2$ is called the equation of the line.

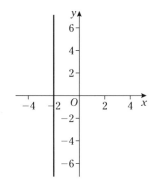

- **A line passes through the origin and has a gradient of $\frac{1}{2}$.**

  The point $P(x, y)$ is on this line if and only if the gradient of OP is $\frac{1}{2}$.

  In terms of x and y, the gradient of OP is $\dfrac{y}{x}$, so the statement above can be written in the form

  $$P(x, y) \text{ is on the line if and only if } \frac{y}{x} = \frac{1}{2}, \quad \text{i.e.} \quad 2y = x$$

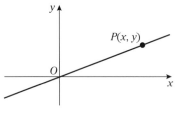

Therefore the coordinates of points on the line satisfy the relationship $2y = x$, and the coordinates of points that are not on the line do not satisfy this relationship.

  $2y = x$ is called the equation of the line.

The equation of a line (straight or curved) is a relationship between the x and y-coordinates of all points on the line that is not satisfied by any other point in the plane.

Examples 3a

1 Find the equation of the line through the points $(1, -2)$ and $(-2, 4)$.

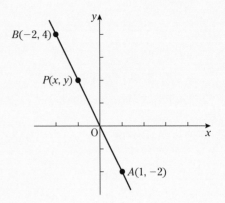

$P(x, y)$ is on the line if and only if the gradient of PA is equal to the gradient of AB (or PB).

The gradient of PA is $\dfrac{y - (-2)}{x - 1} = \dfrac{y + 2}{x - 1}$

The gradient of AB is $\dfrac{-2 - 4}{1 - (-2)} = -2$

Therefore the coordinates of P satisfy the equation $\dfrac{y + 2}{x - 1} = -2$

i.e. $y + 2x = 0$

To find the equation of any line whose gradient is m and that cuts the y-axis at a directed distance c from the origin, we start with

c is called the *intercept on the y-axis.*

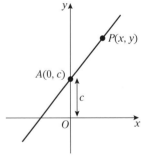

$P(x, y)$ is on this line if and only if the gradient of AP is m.

Therefore the coordinates of P satisfy the equation $\dfrac{y - c}{x - 0} = m$

i.e. $y = mx + c$

This is the *standard form* for the equation of a straight line.

An equation of the form $y = mx + c$ represents a straight line with gradient m and intercept c on the y-axis.

Because the value of m and/or c may be fractional, this equation can be rearranged and expressed as $ax + by + c = 0$, i.e.

$ax + by + c = 0$ where a, b and c are constants, is the equation of a straight line.

In this form c is *not* the intercept.

Examples 3a cont.

2 Write down the gradient of the line $3x - 4y + 2 = 0$ and find the equation of the line through the origin that is perpendicular to the given line.

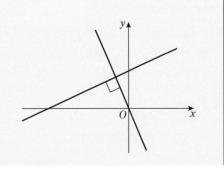

Rearranging $3x - 4y + 2 = 0$ in the standard form gives $y = \frac{3}{4}x + \frac{1}{2}$

Comparing with $y = mx + c$ we can read off the gradient (m) and the intercept on the y-axis.

The gradient of the line is $\frac{3}{4}$.

The gradient of the perpendicular line is $-\dfrac{1}{m}$, i.e. $-\dfrac{4}{3}$ and it passes through the origin so the intercept on the y-axis is 0.

Therefore its equation is $y = -\frac{4}{3}x + 0 \quad \Rightarrow \quad 3y + 4x = 0$

3 Sketch the line $x - 2y + 3 = 0$

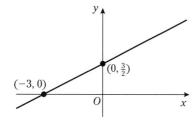

This line can be located accurately in the xy-plane when we know two points on the line. We will use the intercepts on the axes as these can be found easily.

(i.e. $x = 0 \Rightarrow y = \frac{3}{2}$ and $y = 0 \Rightarrow x = -3$)

The diagrams in the worked examples are sketches, not accurate plots, but they show approximately the position of the lines in the plane.

Exercise 3a

1 Write down the equation of the line through the origin and with gradient

(a) 2

(b) -2

(c) $\frac{1}{3}$

(d) $-\frac{1}{4}$

(e) 0

(f) ∞

Draw a sketch showing all these lines on the same set of axes.

2 Write down the equation of the line passing through the origin and perpendicular to

(a) $y = 2x + 3$

(b) $3x + 2y - 4 = 0$

(c) $x - 2y + 3 = 0$

3 Write down the equation of the line passing through $(3, -2)$ and parallel to

(a) $5x - y + 3 = 0$

(b) $x + 7y - 5 = 0$

FINDING THE EQUATION OF A STRAIGHT LINE

Straight lines play a major role in graphical analysis and it is important to be able to find their equations easily.

The equation of a line with gradient m and passing through the point (x_1, y_1)

$P(x, y)$ is a point on the line if and only if the gradient of AP is m

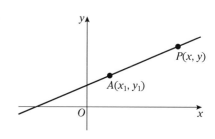

i.e. $\dfrac{y - y_1}{x - x_1} = m$

$\Rightarrow \qquad \boldsymbol{y - y_1 = m(x - x_1)}$ [1]

It is often simpler to work from a diagram than to apply a formula.

The equation of the line passing through (x_1, y_1) and (x_2, y_2)

In the equation $y = mx + c$, m is the gradient of AB, i.e. $m = \dfrac{y_2 - y_1}{x_2 - x_1}$

So the equation of the line through A and B is

$$y - y_1 = \left[\frac{y_2 - y_1}{x_2 - x_1} \right] (x - x_1) \qquad\qquad [2]$$

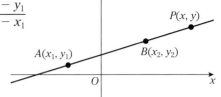

Examples 3b

1 Find the equation of the line with gradient $-\frac{1}{3}$ and passing through $(2, -1)$.

Using [1] with $m = -\frac{1}{3}, x_1 = 2$ and $y_1 = -1$ gives

$$y - (-1) \;\; = -\frac{1}{3}(x - 2)$$

$\Rightarrow \qquad x + 3y + 1 = 0$

Alternatively the equation of this line can be found from the standard form of the equation of a straight line, i.e. $y = mx + c$

Using $y = mx + c$ and $m = -\frac{1}{3}$ we have $y = -\frac{1}{3}x + c$

The point $(2, -1)$ lies on this line so its coordinates satisfy the equation.

i.e. $\qquad -1 = -\frac{1}{3}(2) + c \;\; \Rightarrow \;\; c = -\frac{1}{3}$

Therefore $\qquad y = -\frac{1}{3}x - \frac{1}{3}$

$\Rightarrow \qquad x + 3y + 1 = 0$

2 Find the equation of the line through the points $(1, -2), (3, 5)$.

Using [2] with $x_1 = 1, y_1 = -2, x_2 = 3$ and $y_2 = 5$ gives

$$y - (-2) = \frac{5 - (-2)}{3 - 1}(x - 1)$$

$\Rightarrow \qquad 7x - 2y - 11 = 0$

The worked examples in this book contain a lot of explanation but with practice, you can use any of the methods illustrated to write down the equation of a straight line directly.

Examples 3b cont.

3 Find the equation of the line through $(1, 2)$ that is perpendicular to the line $3x - 7y + 2 = 0$

Expressing $3x - 7y + 2 = 0$ in standard form gives $y = \frac{3}{7}x + \frac{2}{7}$

Hence the given line has gradient $\frac{3}{7}$.

So the required line has a gradient of $-\frac{7}{3}$ and it passes through $(1, 2)$.

Using $y - y_1 = m(x - x_1) \;\; \Rightarrow \;\; y - 2 = \frac{-7}{3}(x - 1)$

$\Rightarrow \qquad 7x + 3y - 13 = 0$

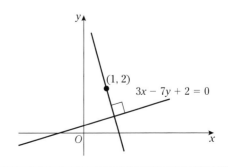

In the last example the line perpendicular to $3x - 7y + 2 = 0$
has equation $\qquad\qquad\qquad 7x + 3y - 13 = 0$

i.e. the coefficients of x and y have been transposed and the sign between the x and y terms has changed. This is a particular example of the general fact that

given a line with equation $ax + by + c = 0$ then the equation of any perpendicular line is
$$bx - ay + k = 0$$

This property of perpendicular lines can be used to shorten the working of problems. For example, to find the equation of the line passing through $(2, -6)$ that is perpendicular to the line $5x - y + 3 = 0$, the required line has an equation $x + 5y + k = 0$. We can use the coordinates $(2, -6)$ which satisfy this equation to find the value of k.

Exercise 3b

1 Find the equation of the line with the given gradient and passing through the given point.

   (a) $3, (4, 9)$ (b) $-5, (2, -4)$

   (c) $\frac{1}{4}, (4, 0)$ (d) $0, (-1, 5)$

   (e) $-\frac{2}{5}, \left(\frac{1}{2}, 4\right)$ (f) $-\frac{3}{8}, \left(\frac{22}{5}, -\frac{5}{2}\right)$

2 Find the equation of the line passing through the points

   (a) $(0, 1), (2, 4)$ (b) $(-1, 2), (1, 5)$

   (c) $(3, -1), (3, 2)$

3 Determine which of the following pairs of lines are perpendicular.

   (a) $x - 2y + 4 = 0$ and $2x + y = 3$

   (b) $x + 3y = 6$ and $3x + y + 2 = 0$

   (c) $x + 3y - 2 = 0$ and $y = 3x + 2$

   (d) $y + 2x + 1 = 0$ and $x = 2y - 4$

4 Find the equation of the line through the point $(5, 2)$ and perpendicular to the line $x - y + 2 = 0$

5 Find the equation of the perpendicular bisector of the line joining

   (a) $(0, 0), (2, 4)$

   (b) $(3, -1), (-5, 2)$

   (c) $(5, -1), (0, 7)$

6 Find the equation of the line through the origin that is parallel to the line $4x + 2y - 5 = 0$

7 The line $4x - 5y + 20 = 0$ cuts the x-axis at A and the y-axis at B. Find the equation of the median through O of $\triangle OAB$.

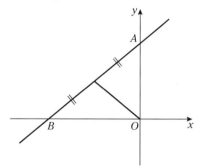

8 Find the equation of the perpendicular through O of the triangle OAB defined in Question 7.

9 Find the equation of the perpendicular height from $(5, 3)$ to the line $2x - y + 4 = 0$

10 The points $A(1, 4)$ and $B(5, 7)$ are two adjacent vertices of a parallelogram $ABCD$. The point $C(7, 10)$ is another vertex of the parallelogram. Find the equation of the side CD.

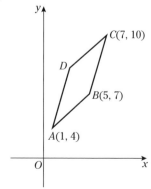

INTERSECTION

The point where two lines (or curves) cut is called a point of intersection.

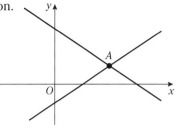

If A is the point of intersection of the lines $y - 3x + 1 = 0$ [1]
and $y + x - 2 = 0$ [2]

then the coordinates of A satisfy both of these equations.
A can be found by solving [1] and [2] simultaneously, i.e.

$[2] - [1]$ $\Rightarrow$ $4x - 3 = 0$ $\Rightarrow$ $x = \frac{3}{4}$ and $y = \frac{5}{4}$

Therefore $\left(\frac{3}{4}, \frac{5}{4}\right)$ is the point of intersection.

Example 3c

A circle has radius 4 and its centre is the point $C(5, 3)$.

(a) Show that the points $A(5, -1)$ and $B(1, 3)$ are on the circumference of the circle.

(b) Prove that the perpendicular bisector of AB goes through the centre of the circle.

(a) From the diagram, $BC = 4$

$\therefore$ B is on the circumference.

Similarly $AC = 4$

$\therefore$ A is on the circumference.

(b) The midpoint, M, of AB is $\left[\dfrac{5 + 1}{2}, \dfrac{-1 + 3}{2}\right]$ i.e. $(3, 1)$

The gradient of AB is $\dfrac{-1 - 3}{5 - 1} = -1$

If l is the perpendicular bisector of AB, its gradient is 1 and it goes through $(3, 1)$.

$\therefore$ the equation of l is $y - 1 = 1(x - 3)$ $\Rightarrow$ $y = x - 2$ [1]

In equation [1], when $x = 5$, $y = 3$

$\therefore$ the point $(5, 3)$ is on l.

i.e. the perpendicular bisector of AB goes through C.

Exercise 3c

1 Show that the triangle whose vertices are (1, 1), (3, 2) and (2, −1) is isosceles.

2 Find the area of the triangular region enclosed by the x and y-axes and the line $2x − y − 1 = 0$

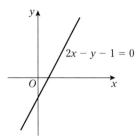

3 Find the coordinates of the triangular region enclosed by the lines $y = 0$, $y = x + 5$ and $x + 2y − 6 = 0$

4 Write down the equation of the perpendicular bisector of the line joining the points $(2, −3)$ and $\left(-\frac{1}{2}, 3\frac{1}{2}\right)$.

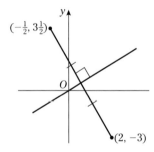

5 Find the equation of the line through $A(5, 2)$ that is perpendicular to the line $y = 3x − 5$. Hence find the coordinates of the foot of the perpendicular from A to the line.

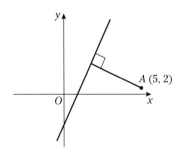

6 The coordinates of a point P are $(t + 1, 2t − 1)$. Sketch the position of P when $t = −1, 0, 1$ and 2. Show that these points are collinear and write down the equation of the line on which they lie.

7 The centre of a circle is at the point $C(3, 7)$ and the point $A(5, 3)$ is on the circumference of the circle. Find

   (a)   the radius of the circle

   (b)   the equation of the line through A that is perpendicular to AC.

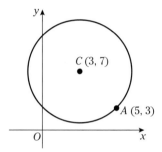

8 The equations of two sides of a square are $y = 3x − 1$ and $x + 3y − 6 = 0$. $(0, −1)$ is one vertex of the square. Find the coordinates of the other vertices.

9 The lines $y = 2x$, $2x + y − 12 = 0$ and $y = 2$ enclose a triangular region of the xy-plane. Find

   (a)   the coordinates of the vertices of this region

   (b)   the area of this region.

4 Circular measure

After studying this chapter you should be able to

- understand the definition of a radian, and use the relationship between radians and degrees
- use the formulae $s = r\theta$ and $A = \frac{1}{2}r^2\theta$ in solving problems concerning the arc length and sector area of a circle.

THE RADIAN

An angle is a measure of rotation and the units used so far are the revolution and the degree.

There is another unit called the radian. This unit simplifies work in circles and in trigonometry and it is used in most further work in mathematics.

O is the centre of a circle and an arc PQ is drawn so that its length is equal to the radius of the circle. The angle POQ is called a *radian* (one radian is written 1 rad or 1^c).

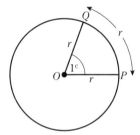

An arc equal in length to the radius of a circle subtends an angle of 1 radian at the centre.

'Subtends an angle' means the angle enclosed by lines from the ends of the arc to the centre of the circle.

It follows that the number of radians in a complete revolution is the number of times the radius divides into the circumference.

The circumference of a circle is of length $2\pi r$, so the number of radians in a revolution is $2\pi r \div r$ which is 2π

i.e. $\qquad 2\pi$ radians $= 360°$

Further, π radians $= 180°$ and $\frac{1}{2}\pi$ radians $= 90°$

$\frac{1}{2}\pi$ can be written as $\dfrac{\pi}{2}$, $\frac{2}{3}\pi$ as $\dfrac{2\pi}{3}$, and so on.

When an angle is given in terms of π we usually omit the radian symbol, i.e. we write $180° = \pi$ (not $180° = \pi$ rad).

If an angle is a simple fraction of $180°$, it can easily be given in terms of π

e.g. $\qquad 60° = \frac{1}{3}$ of $180° = \frac{1}{3}\pi = \dfrac{\pi}{3}$ and $135° = \frac{3}{4}$ of $180° = \frac{3}{4}\pi = \dfrac{3\pi}{4}$

Conversely, $\qquad \dfrac{7\pi}{6} = \frac{7}{6}\pi = \frac{7}{6}$ of $180° = 210°$

and $\qquad \frac{2}{3}\pi = \frac{2}{3}$ of $180° = 120°$

Angles that are not simple fractions of 180°, or π, can be converted by using the relationship $\pi = 180°$, taking the value of π from a calculator,

e.g. $73° = \frac{73}{180} \times \pi = 1.27 \, \text{rad}$ (correct to 3 s.f.)

and $2.36 \, \text{rad} = \frac{2.36}{\pi} \times 180° = 135°$ (correct to the nearest degree)

Now $1 \, \text{rad} = \frac{1}{\pi} \times 180° = 57°$ (correct to 2 s.f.),

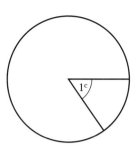

i.e. 1 radian is just a little less than 60°. This helps to 'see' the size of a radian.

Examples 4a

1 Express 75° in radians in terms of π.

$$75° = 75 \times \frac{\pi}{180} \text{ radians} = \frac{5\pi}{12} \text{ radians}$$

$180° = \pi \text{ radians} \Rightarrow 1° = \frac{\pi}{180} \text{ radians}$; so to convert degrees to radians, multiply by $\frac{\pi}{180}$

2 Express $\frac{1}{16}\pi$ radians in degrees.

$$\frac{1}{16}\pi \text{ radians} = \frac{\pi}{16} \times \frac{180°}{\pi} = \frac{45°}{4} = 11\frac{1}{4}°$$

$\pi \text{ radians} = 180° \Rightarrow 1^c = \frac{180}{\pi}$, so to convert radians to degrees, multiply by $\frac{180}{\pi}$

Exercise 4a

1 Express each of the following angles in radians as a fraction of π.

(a) 45°   (b) 150°   (c) 30°

(d) 90°   (e) 270°   (f) 120°

2 Express each of the following angles in degrees.

(a) $\frac{1}{6}\pi$ (b) π (c) $\frac{1}{10}\pi$

(d) $\frac{\pi}{3}$ (e) $\frac{5}{6}\pi$ (f) $\frac{1}{12}\pi$

(g) $\frac{7\pi}{6}$ (h) $\frac{3}{4}\pi$ (i) $\frac{\pi}{9}$

(j) $\frac{3}{2}\pi$ (k) $\frac{4}{9}\pi$ (l) $\frac{1}{4}\pi$

(m) $\frac{3\pi}{5}$ (n) $\frac{1}{8}\pi$

3 Express each of the following angles in radians correct to 2 decimal places.

(a) 35°   (b) 47.2°   (c) 93°

(d) 233°   (e) 14.1°   (f) 117°

(g) 370°

4 Express each of the following angles in degrees correct to 1 decimal place.

(a) 1.7 rad

(b) 3.32 rad

(c) 1 rad

(d) 2.09 rad

(e) 5 rad

(f) 6.28319 rad

5 Use your calculator to find

(a) sin 1.2 rad

(b) cos 0.35 rad

(c) tan 1.47 rad

(d) cos 2.5 rad

There is no need to change the angle to degrees: set the angle mode on your calculator to radians, then $\sin \theta \, \text{rad}$ can be keyed in directly. Similarly with the mode in radians, a calculator will give the angle in radians for which, say, $\sin \theta = 0.7$

THE LENGTH OF AN ARC

From the definition of a radian, the arc that subtends an angle of 1 radian at the centre of the circle is of length r. Therefore if an arc subtends an angle of θ radians at the centre, the length of the arc is $r\theta$.

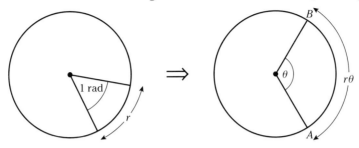

The length of arc $AB = r\theta$

Examples 4b

1 An arc subtends an angle of $\dfrac{\pi}{3}$ at the centre of a circle with radius 4.5 cm.

Find the length of the arc in terms of π.

Length of arc $= r\theta$

$$= 4.5 \times \frac{\pi}{3} = 1.5\pi$$

2 Find, in radians, the angle subtended at the centre of a circle of radius 4 cm by an arc of length 0.8 cm.

Length of arc $= r\theta$

$\therefore \qquad 0.8 = 4\theta \qquad \Rightarrow \qquad \theta = 0.2\,\text{rad}$

3 AB and AC are tangents to a circle of radius 5 cm and centre P. $AP = 10$ cm as shown in the diagram. Find the length of the major arc BC.

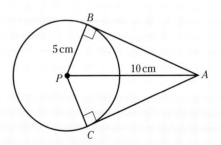

In $\triangle ABP$, $BP = 5$ cm, $AP = 10$ cm and $\angle ABP = 90°$

The tangents AB and AC are perpendicular to the radii PB and PC respectively.

Therefore $\qquad \cos APB = \frac{5}{10} = \frac{1}{2}$

$\Rightarrow \qquad\qquad \angle APB = 60° = \frac{1}{3}\pi$

Similarly $\qquad \angle APC = \frac{1}{3}\pi$

$\therefore \qquad\qquad \angle BPC = \frac{2}{3}\pi$

Hence the angle subtended at P by the major arc BC is $2\pi - \frac{2}{3}\pi = \frac{4}{3}\pi$

The length of major arc BC is $5 \times \frac{4}{3}\pi = 20.9$ cm (3 s.f.) $\qquad$ Using length of arc $= r\theta$

Exercise 4b

1 Find, in terms of π, the length of the arc that subtends an angle of $\frac{1}{6}\pi$ radians at the centre of a circle of radius 4 cm.

2 An arc subtends an angle of $\frac{5}{4}\pi$ radians at the centre of a circle of radius 10 cm. Find, in terms of π, the length of the arc.

3 Find, in radians, the angle subtended at the centre of a circle of radius 5 cm by an arc of length 12 cm.

4 Find the size of the angle subtended at the centre of a circle of radius 65 mm by an arc of length 45 mm. Give your answer in radians.

5 Find the radius of a circle in which an arc of length 15 cm subtends an angle of π radians at the centre.

6 An arc of length 20 cm subtends an angle of $\frac{4}{5}\pi$ radians at the centre of a circle. Find the radius of the circle.

7 Find, in terms of π, the length of the arc that subtends an angle of $60°$ at the centre of a circle of radius 12 cm.

8 An arc of length 15 cm subtends an angle of $45°$ at the centre of a circle. Find, in terms of π, the radius of the circle.

9

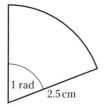

A sector of a circle of radius 2.5 cm subtends an angle of 1 radian at the centre. Find the perimeter of the sector.

10

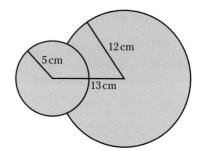

Two circles of radii 5 cm and 12 cm overlap so that the distance between their centres is 13 cm.
Find the perimeter of the shape.

THE AREA OF A SECTOR

We know that $\dfrac{\text{area of sector}}{\text{area of circle}} = \dfrac{\text{angle contained in the sector}}{\text{complete angle at the centre}}$

A sector contains an angle of θ radians at the centre of the circle.

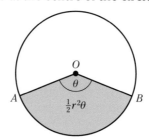

The complete angle at the centre of the circle is 2π, hence

$$\frac{\text{area of sector}}{\text{area of circle}} = \frac{\theta}{2\pi}$$

The area of the circle is πr^2

$\Rightarrow$ $\qquad$ area of sector $= \dfrac{\theta}{2\pi} \times \pi r^2 = \frac{1}{2}r^2\theta$

The area of sector $AOB = \frac{1}{2}r^2\theta$

Examples 4c

1 Find, in terms of π, the area of the sector of a circle of radius 3 cm that contains an angle of $\frac{\pi}{5}$.

Area of sector $= \frac{1}{2}r^2\theta = \frac{1}{2}(3)^2\left(\frac{\pi}{5}\right)$ cm$^2 = \frac{9\pi}{10}$ cm^2

2 AB is a chord of a circle with centre O and radius 4 cm. AB is of length 4 cm and divides the circle into two segments.
Find the area of the minor segment.

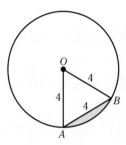

ABC is an equilateral triangle, so each angle is $60°$, i.e. $\frac{\pi}{3}$ rad.

Area of sector $AOB = \frac{1}{2}r^2\theta = \frac{1}{2}(4^2)\left(\frac{1}{3}\pi\right) = 8.3775...$

Area of $\Delta AOB = \frac{1}{2}r^2\sin\theta = \frac{1}{2}(4)(4)(\sin 60°) = 6.9282...$

To use the formula for the area of a sector, the angle must be in radians. To find the area of the minor segment, we subtract the area of ΔAOB from the area of sector AOB.

Area of minor segment $=$ area of sector $AOB -$ area of ΔAOB

$= 8.3775... - 6.9282... = 1.449...$

The area of the minor segment is 1.45 cm^2 (3 s.f.)

Exercise 4c

1 A sector of a circle of radius 4 cm contains an angle of $30°$. Find the area of the sector.

2 A sector of a circle of radius 8 cm contains an angle of $135°$. Find the area of the sector.

3 The area of a sector of a circle of radius 2 cm is π cm^2. Find the angle contained by the sector.

4 The area of a sector of a circle of radius 5 cm is 12 cm^2. Find the angle contained by the sector.

5 A sector of a circle of radius 10 cm contains an angle of $\frac{5\pi}{6}$.
Find the area of the sector.

6 An arc of length 15 cm subtends an angle π at the centre of a circle. Find the radius of the circle and hence the area of the sector containing the angle π.

7 A sector of a circle has an area 3π cm^2 and contains an angle $\frac{1}{6}\pi$. Find the radius of the circle.

8 A sector of a circle has an area 6π cm^2 and contains an angle of $45°$. Find the radius of the circle.

9

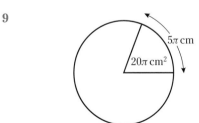

An arc of a circle is of length 5π cm and the sector it bounds has an area of 20π cm^2. Find the radius of the circle.

10 Calculate, in radians, the angle at the centre of a circle of radius 8.3 cm contained in a sector of area 9.74 cm^2.

11 In a circle with centre O and radius 5 cm, AB is a chord of length 8 cm. Find

    (a) the area of $\triangle AOB$

    (b) the area of the sector AOB.

12 A chord of length 1 cm divides a circle of radius 0.7 cm into two segments. Find the area of each segment.

13 A chord PQ, of length 12.6 cm, subtends an angle of $\frac{2}{3}\pi$ at the centre of a circle. Find

    (a) the length of the arc PQ

    (b) the area of the minor segment cut off by the chord PQ.

14 Two circles, each of radius 14 cm, are drawn with their centres 20 cm apart. Find the length of their common chord. Find also the area common to the two circles.

MIXED PROBLEMS

Example 4d

The diagram shows a circle centre O, radius r.

$\angle AOB = \theta$ radians

The area of $\triangle AOB$ is twice the area of the shaded segment.

Show that $\theta = \frac{3}{2}\sin\theta$

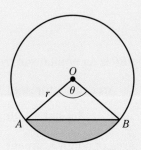

The area of $\triangle AOB = \frac{1}{2}r^2\sin\theta$

The area of sector $AOB = \frac{1}{2}r^2\theta$

Area of $\triangle AOB = 2 \times$ area of the shaded segment is given, so use this to find a relationship between the expressions for these areas.

Therefore the area of the shaded segment is equal to $\frac{1}{2}r^2\theta - \frac{1}{2}r^2\sin\theta$

Therefore $\quad \frac{1}{2}r^2\sin\theta = 2 \times \left(\frac{1}{2}r^2\theta - \frac{1}{2}r^2\sin\theta\right)$

This can be simplified by multiplying both sides by 2 and dividing both sides by r^2.

$\Rightarrow \qquad \sin\theta = 2\theta - 2\sin\theta$

$\Rightarrow \qquad 3\sin\theta = 2\theta$

i.e. $\qquad \theta = \frac{3}{2}\sin\theta$

Exercise 4d

1 A chord of a circle subtends an angle of θ radians at the centre of the circle. The area of the minor segment cut off by the chord is one-eighth of the area of the circle. Prove that $4\theta = \pi + 4\sin\theta$

2

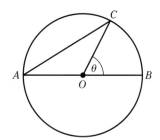

AB is the diameter of a circle, centre O. C is a point on the circumference such that $\angle COB = \theta$ radians. The area of the minor segment cut off by AC is equal to twice the area of the sector BOC. Show that $3\theta = \pi - \sin\theta$

3

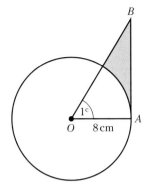

The diagram shows a sector of a circle, centre O, containing an angle of 1 radian. Find the area of the shaded region of the diagram.

4

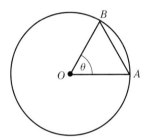

O is the centre of a circle of radius r cm. A chord AB subtends an angle of θ radians at O.

(a) Show that the area of the minor segment cut off by AB is equal to

$$\tfrac{1}{2}r^2(\theta - \sin \theta)$$

(b) The area of the circle is 20 times the area of the minor segment. Show that

$$\sin \theta = \theta - \frac{\pi}{10}$$

5

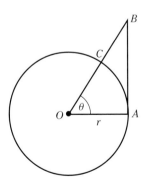

AB is a tangent to the circle centre O. $\angle AOC = \theta$ radians. Show that the perimeter of the section bounded by the lines AB, BC and the arc AC is given by

$$r\left(\tan \theta + \frac{1}{\cos \theta} + \theta - 1 \right)$$

6

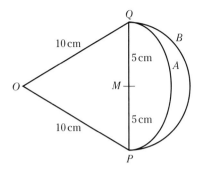

The diagram shows two arcs, A and B. Arc A is part of the circle, centre O and radius OP. Arc B is part of the circle, centre M and radius PM, where M is the midpoint of PQ.

(a) Write down the angle POQ in radians.

(b) Show that the area enclosed by the two arcs is equal to

$$25 \left(\sqrt{3} - \frac{\pi}{6} \right). \qquad \left[\text{Use } \sin \frac{\pi}{3} = \frac{\sqrt{3}}{2} \right]$$

7

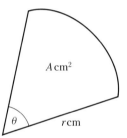

The diagram shows a sector of a circle of radius r cm containing an angle of θ radians. The area of the sector is A cm^2 and the perimeter of the sector is 50 cm.

(a) Find θ in terms of r.

(b) Show that $A = 25r - r^2$

8

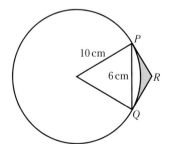

A chord PQ of length 6 cm is drawn in a circle of radius 10 cm. The tangents to the circle at P and Q meet at R. Find the area enclosed by PR, QR and the minor arc PQ.

9

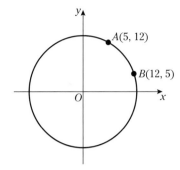

The diagram shows a circle with radius 13 cm whose centre is at the origin. The points $A(5, 12)$ and $B(12, 5)$ are on the circumference of the circle. Find the length of the arc AB.

10

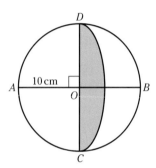

The diagram shows a circle of radius 10 cm. AB and CD are perpendicular diameters of the circle. The arc CD is an arc of the circle centre A and radius AD. Find the area of the shaded region.

11

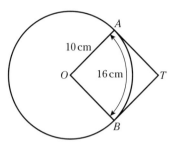

AT and BT are tangents to a circle, centre O and radius 10 cm. The length of the arc AB is 16 cm. Find

(a)   the size of $\angle AOB$

(b)   the area of $\triangle ABT$.

12

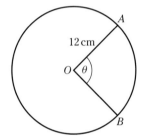

The minor sector AOB of the circle contains an angle of θ rad. The perimeter of the minor sector is 60 cm. Find the area of the major sector.

Summary 1

QUADRATIC EQUATIONS

The general quadratic equation is $ax^2 + bx + c = 0$

The roots of this equation can be found by factorising when this is possible, or completing the square, or by using the formula

$$x = \frac{-b \pm \sqrt{b^2 - 4ac}}{2a}$$

When $b^2 - 4ac > 0$, the roots are real and different.

When $b^2 - 4ac = 0$, the roots are real and equal.

When $b^2 - 4ac < 0$, the roots are not real.

COORDINATE GEOMETRY

Length of AB is $\sqrt{(x_2 - x_1)^2 + (y_2 - y_1)^2}$

Midpoint, M, of AB is $\left[\frac{1}{2}(x_1 + x_2), \frac{1}{2}(y_1 + y_2)\right]$

Gradient of AB is $\dfrac{y_2 - y_1}{x_2 - x_1}$

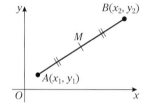

Parallel lines have equal gradients. When two lines are perpendicular the product of their gradient is -1.

The standard equation of a straight line is $y = mx + c$, where m is its gradient and c its intercept on the y-axis.

Any equation of the form $ax + by + c = 0$ gives a straight line.

The equation of a line passing through (x_1, y_1) and with gradient m is

$$y - y_1 = m(x - x_1)$$

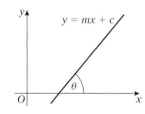

Given a line with equation $ax + by + c = 0$, then any perpendicular line has equation $bx - ay + k = 0$

CIRCULAR MEASURE

One radian (1^c) is the size of the angle subtended at the centre of a circle by an arc equal in length to the radius of the circle.

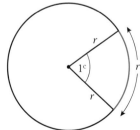

The length of arc AB is $r\theta$.

The area of sector AOB is $\frac{1}{2}r^2\theta$.

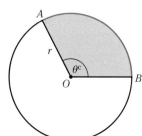

Summary exercise 1

1

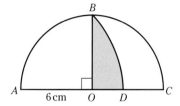

The diagram shows a semicircle *ABC* with centre *O* and radius 6 cm. The point *B* is such that angle *BOA* is 90° and *BD* is an arc of a circle with centre *A*. Find

(i) the length of the arc *BD* [4]

(ii) the area of the shaded region. [3]

Cambridge, Paper 11 Q5 N09

2

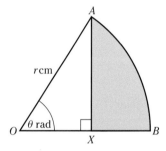

In the diagram, *AB* is an arc of a circle, centre *O* and radius *r* cm, and angle *AOB* = θ radians. The point *X* lies on *OB* and *AX* is perpendicular to *OB*.

(i) Show that the area, *A* cm², of the shaded region *AXB* is given by

$$A = \tfrac{1}{2}r^2\,(\theta - \sin\theta\cos\theta)$$ [3]

(ii) In the case where *r* = 12 and $\theta = \tfrac{1}{6}\pi$, find the perimeter of the shaded region *AXB*, leaving your answer in terms of $\sqrt{3}$ and π. [4]

Cambridge, Paper 1 Q7 N07

3

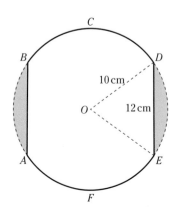

The diagram shows a metal plate *ABCDEF* which has been made by removing the two shaded regions from a circle of radius 10 cm and centre *O*. The parallel edges *AB* and *ED* are both of length 12 cm.

(i) Show that angle *DOE* is 1.287 radians, correct to 4 significant figures. [2]

(ii) Find the perimeter of the metal plate. [3]

(iii) Find the area of the metal plate. [3]

Cambridge, Paper 13 Q7 J10

4

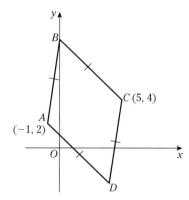

The diagram shows a rhombus *ABCD* in which the point *A* is (−1, 2), the point *C* is (5, 4) and the point *B* lies on the *y*-axis. Find

(i) the equation of the perpendicular bisector of *AC* [3]

(ii) the coordinates of *B* and *D* [3]

(iii) the area of the rhombus. [3]

Cambridge, Paper 13 Q8 J10

5

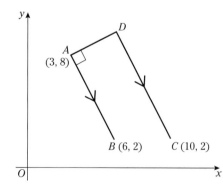

The three points *A*(3, 8), *B*(6, 2) and *C*(10, 2) are shown in the diagram. The point *D* is such that the line *DA* is perpendicular to *AB* and *DC* is parallel to *AB*. Calculate the coordinates of *D*. [7]

Cambridge, Paper 1 Q6 N07

6

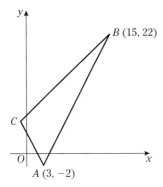

The diagram shows a triangle ABC in which A is $(3, -2)$ and B is $(15, 22)$. The gradients of AB, AC and BC are $2m$, $-2m$ and m respectively, where m is a positive constant.

(i) Find the gradient of AB and deduce the value of m. [2]

(ii) Find the coordinates of C. [4]

The perpendicular bisector of AB meets BC at D.

(iii) Find the coordinates of D. [4]

Cambridge, Paper 11 Q8 J10

7

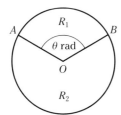

The diagram shows a circle with centre O. The circle is divided into two regions, R_1 and R_2, by the radii OA and OB, where angle $AOB = \theta$ radians. The perimeter of the region R_1 is equal to the length of the major arc AB.

(i) Show that $\theta = \pi - 1$ [3]

(ii) Given that the area of region R_1 is $30\,\text{cm}^2$, find the area of region R_2, correct to 3 significant figures. [4]

Cambridge, Paper 1 Q5 J09

8

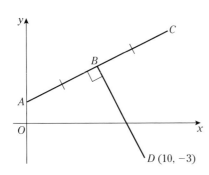

The diagram shows points A, B and C lying on the line $2y = x + 4$. The point A lies on the y-axis and $AB = BC$. The line from $D(10, -3)$ to B is perpendicular to AC. Calculate the coordinates of B and C. [7]

Cambridge, Paper 1 Q8 J09

9

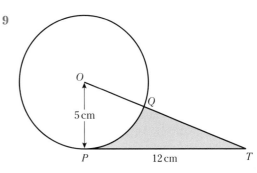

The diagram shows a circle with centre O and radius $5\,\text{cm}$. The point P lies on the circle, PT is a tangent to the circle and $PT = 12\,\text{cm}$. The line OT cuts the circle at the point Q.

(i) Find the perimeter of the shaded region. [4]

(ii) Find the area of the shaded region. [3]

Cambridge, Paper 1 Q5 part one J08

10

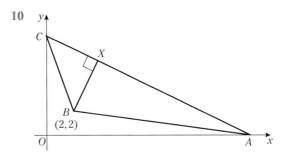

In the diagram, the points A and C lie on the x- and y-axes respectively and the equation of AC is $2y + x = 16$. The point B has coordinates $(2, 2)$. The perpendicular from B to AC meets AC at the point X.

(i) Find the coordinates of X. [4]

The point D is such that the quadrilateral $ABCD$ has AC as a line of symmetry.

(ii) Find the coordinates of D. [2]

(iii) Find, correct to 1 decimal place, the perimeter of $ABCD$. [3]

Cambridge, Paper 1 Q11 J08

11

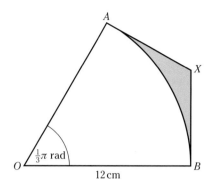

In the diagram, OAB is a sector of a circle with centre O and radius 12 cm. The lines AX and BX are tangents to the circle at A and B respectively. Angle $AOB = \frac{1}{3}\pi$ radians.

(i) Find the exact length of AX, giving your answer in terms of $\sqrt{3}$. [2]

(ii) Find the area of the shaded region, giving your answer in terms of π and $\sqrt{3}$. [3]

Cambridge, Paper 1 Q5 J07

12

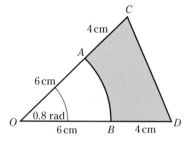

In the diagram, OCD is an isosceles triangle with $OC = OD = 10$ cm and angle $COD = 0.8$ radians. The points A and B, on OC and OD respectively, are joined by an arc of a circle with centre O and radius 6 cm. Find

(i) the area of the shaded region [3]

(ii) the perimeter of the shaded region. [4]

Cambridge, Paper 1 Q5 J04

13 The curve $y = 9 - \dfrac{6}{x}$ and the line $y + x = 8$ intersect at two points. Find

(i) the coordinates of the two points [4]

(ii) the equation of the perpendicular bisector of the line joining the two points. [4]

Cambridge, Paper 1 Q6 J04

5 Functions

After studying this chapter you should be able to

- understand the terms function, domain, range, one−one function, inverse function and composition of functions
- identify the range of a given function in simple cases, and find the composition of two given functions
- determine whether or not a given function is one−one, and find the inverse of a one−one function in simple cases
- illustrate in graphical terms the relation between a one−one function and its inverse
- carry out the process of completing the square for a quadratic polynomial $ax^2 + bx + c$, and use this form, e.g. to locate the vertex of the graph of $y = ax^2 + bx + c$ or to sketch the graph.

MAPPINGS

When the number 2 is entered in a calculator and then the x^2 button is pressed, the display shows the number 4.

2 is mapped to 4, which is denoted by $2 \mapsto 4$

Under this rule, that is, squaring the input number,

$3 \mapsto 9$, $25 \mapsto 625$, $0.2 \mapsto 0.04$, $-2 \mapsto 4$ and (any real number) $\mapsto$ (the square of that number)

This is denoted by $x \mapsto x^2$, for $x \in \mathbb{R}$ $x \in \mathbb{R}$ means x is any real number.

This mapping can be represented graphically by plotting values of x^2 against values of x.

The graph, and knowledge of what happens when we square a number, show that one input number gives just one output number.

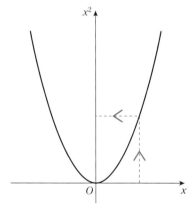

But the mapping that maps a number to its square root, e.g. $4 \mapsto -2$ and 2 gives a real output only when the input number is greater than or equal to zero (negative numbers do not have real square roots).

This mapping can be written as $x \mapsto \pm\sqrt{x}$ for $x \geqslant 0$, $x \in \mathbb{R}$

The graphical representation of this mapping shows that one input value gives two output values.

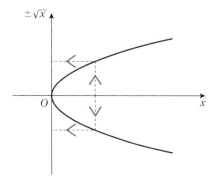

FUNCTIONS

For the mapping $x \mapsto x^2$, for $x \in \mathbb{R}$, one input number gives one output number.

The mapping $x \mapsto \pm\sqrt{x}$ for $x \geqslant 0$, $x \in \mathbb{R}$ gives two outputs for every one input number.
The word function is used for any mapping where one input value gives one output value.

A function is a rule that maps a single number to another single number for a defined set of input numbers.

The mapping $x \mapsto \pm\sqrt{x}$ for $x \geqslant 0$, $x \in \mathbb{R}$ is not a function because it does not satisfy this condition.

Using f for function and the symbol : to mean 'such that' we write $f : x \mapsto x^2$ for $x \in \mathbb{R}$ to mean f is the function that maps x to x^2 for all real values of x.

Example 5a

Determine whether these mappings are functions, for $x \in \mathbb{R}$

(a) $x \mapsto \dfrac{1}{x}$ 　　　　　　　　　　　　　　　　(b) $x \mapsto y$ where $y^2 - x = 0$

(a) For any value of x, except $x = 0$, $\dfrac{1}{x}$ has a single value,

therefore $x \mapsto \dfrac{1}{x}$ is a function provided that $x = 0$ is excluded.

$\dfrac{1}{0}$ is meaningless, so to make this mapping a function we have to exclude 0 as an input value.
The function can be described by $f : x \mapsto \dfrac{1}{x}, x \neq 0, x \in \mathbb{R}$

(b) If, as an example, we input $x = 4$, then the output is the value of y given by $y^2 - 4 = 0$
i.e. $y = 2$ and $y = -2$

Therefore an input gives more than one value for the output, so $x \mapsto y$ where $y^2 - x = 0$ is not a function.

Exercise 5a

Determine which of these mappings are functions.

1 $x \mapsto 2x - 1, x \in \mathbb{R}$

2 $x \mapsto x^3 + 3, x \in \mathbb{R}$

3 $x \mapsto \dfrac{1}{x} - 1, x \in \mathbb{R}$

4 $x \mapsto t$ where $t^2 = x, x \in \mathbb{R}$

5 $x \mapsto \sqrt{x}, x \in \mathbb{R}$

6 $x \mapsto$ the length of the line from the origin to $(0, x), x \in \mathbb{R}$

DOMAIN AND RANGE

We have assumed that we can use any real number as an input for a function unless some particular numbers have to be excluded because they do not give real numbers as output.

The set of inputs for a function is called the *domain* of the function.

The domain does not have to contain all possible inputs; it can be as wide, or as restricted, as we choose to make it. Hence to define a function fully, the domain must be stated.

If the domain is not stated, we assume that it is the set of all real numbers ($\mathbb{R}$).

The mapping $x \mapsto x^2 + 3$ can be used to define a function f over any domain we choose. Some examples, together with their graphs, are given.

1 $f : x \mapsto x^2 + 3$ for $x \in \mathbb{R}$

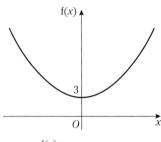

2 $f : x \mapsto x^2 + 3$ for $x \geqslant 0$

Note that the point on the curve where $x = 0$ is included, and we denote this on the curve by a solid circle.

If the domain were $x > 0$, then the point would not be part of the curve, and we indicate this fact by using an open circle.

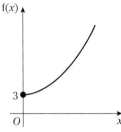

3 $f : x \mapsto x^2 + 3$ for $x = 1, 2, 3, 4, 5$

This time the graphical representation consists of just five discrete (i.e. separate) points.

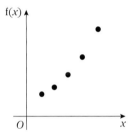

These three examples are not the same function, each is a different function.

For each domain, there is a corresponding set of output numbers.

The set of output numbers is called the *range* of the function.

The notation $f(x)$ represents the output values of a function,

so for $f : x \mapsto x^2$ for $x \in \mathbb{R}$ we have $f(x) = x^2$

For the function defined in (**1**) above, the range is $f(x) \geqslant 3$ and for the function given in (**2**), the range is also $f(x) \geqslant 3$. For the function defined in (**3**), the range is the set of numbers 4, 7, 12, 19, 28.

Examples 5b

1 $f(x) = 2x^2 - 5$, $x \in \mathbb{R}$. Find $f(3)$ and $f(-1)$.

As $f(x)$ is the output of the mapping, $f(3)$ is the output when 3 is the input, i.e. $f(3)$ is the value of $2x^2 - 5$ when $x = 3$

$f(3) = 2(3)^2 - 5 = 13$

$f(-1) = 2(-1)^2 - 5 = -3$

Sometimes a function can be made up from more than one mapping, where each mapping is defined for a different domain. This is illustrated in the next worked example.

Examples 5b cont.

2 The function, f, is defined by $f(x) = x^2$ for $x \leqslant 0$
and $f(x) = x$ for $x > 0$

(a) Find f(4) and f(−4).   (b) Sketch the graph of f.   (c) Give the range of f.

(a)   For $x > 0$, $f(x) = x$

$\therefore$ $f(4) = 4$

For $x \leqslant 0$, $f(x) = x^2$

$\therefore$ $f(-4) = (-4)^2 = 16$

(b) To sketch the graph of a function, we can use what we know about lines and curves in the xy-plane. In this way we can interpret $f(x) = x$ for $x > 0$, as that part of the line $y = x$ which corresponds to positive values of x, and $f(x) = x^2$ for $x \leqslant 0$ as the part of the curve $y = x^2$ that corresponds to negative values of x.

(c) The range of f is $f(x) \geqslant 0$

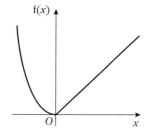

Exercise 5b

1 If $f(x) = 5x - 4$, $x \in \mathbb{R}$, find f(0), f(−4).

2 If $f(x) = 3x^2 + 25$, $x \in \mathbb{R}$, find f(0), f(8).

3 If $f(x) =$ the value of x correct to the nearest integer, $x \in \mathbb{R}$, find f(1.25), f(−3.5), f(12.49).

4 If $f(x) = \sin x$, $x \in \mathbb{R}$, find $f\left(\frac{1}{2}\pi\right)$, $f\left(\frac{2}{3}\pi\right)$.

5 Find the range of each of the following functions.

(a) $f(x) = 2x - 3$ for $x \geqslant 0$, $x \in \mathbb{R}$

(b) $f(x) = x^2 - 5$ for $x \leqslant 0$, $x \in \mathbb{R}$

(c) $f(x) = 1 - x$ for $x \leqslant 1$, $x \in \mathbb{R}$

(d) $f(x) = \dfrac{1}{x}$ for $x \geqslant 2$, $x \in \mathbb{R}$

6 The function f is such that

$f(x) = -x$ for $x < 0$, $x \in \mathbb{R}$

and $f(x) = x$ for $x \geqslant 0$, $x \in \mathbb{R}$

(a) Find the value of f(5), f(−4), f(−2) and f(0).

(b) Sketch the graph of the function.

7 The function f is such that

$f(x) = x$ for $0 \leqslant x \leqslant 5$, $x \in \mathbb{R}$

and $f(x) = 5$ for $x > 5$, $x \in \mathbb{R}$

(a) Find the value of f(0), f(2), f(4), f(5) and f(7).

(b) Sketch the graph of the function.

(c) Give the range of the function.

CURVE SKETCHING

When functions have similar definitions they usually have common properties and graphs of the same form. When the common properties of a group of functions are known, the graph of any one member of the group can be sketched without having to plot points.

Quadratic functions

The general form of a quadratic function is

$$f(x) = ax^2 + bx + c \text{ for } x \in \mathbb{R}$$

where a, b and c are constants and $a \neq 0$

When the graphs of quadratic functions for a variety of values of a, b and c are drawn, the basic shape of the curve is always the same. This shape is called a *parabola*.

Every parabola has an axis of symmetry that goes through the vertex, i.e. the point where the curve turns back upon itself.

If the coefficient of x^2 is positive, i.e. $a > 0$, then f(x) has a least value, and the parabola looks like this.

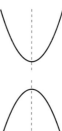

If the coefficient of x^2 is negative, i.e. $a < 0$, then f(x) has a greatest value, and the parabola is this way up.

Examples 5c

1 Express $2x^2 - 7x - 4$ in the form $a(x - b)^2 - c$ where a, b and c are positive constants.

Find the greatest or least value of the function given by f$(x) = 2x^2 - 7x - 4$, $x \in \mathbb{R}$, and hence sketch the graph of f(x).

The simplest way to do this is to expand $a(x - b)^2 - c$ and compare the coefficients of x^2 and x and the constants.

$$2x^2 - 7x - 4 = a(x^2 - 2bx + b^2) - c$$
$$= ax^2 - 2abx + ab^2 - c$$
$$\Rightarrow \qquad a = 2$$
$$-2ab = -7 \text{ so } b = \tfrac{7}{4}$$
$$ab^2 - c = -4 \text{ so } c = \tfrac{49}{8} + 4 = \tfrac{81}{8}$$
$$\therefore \qquad 2x^2 - 7x - 4 = 2(x - \tfrac{7}{4})^2 - \tfrac{81}{8}$$

Alternatively, $2x^2 - 7x - 4 = 2\{(x^2 - \tfrac{7}{2}x) - 2\}$
$$= 2\{(x^2 - \tfrac{7}{2}x + \tfrac{49}{16}) - 2 - \tfrac{49}{16}\}$$

completing the square on $x^2 - \tfrac{7}{2}x$ by adding $\tfrac{49}{16}$ then subtracting it
$$= 2(x - \tfrac{7}{4})^2 - \tfrac{81}{8}$$

f(x) has a least value of $-\tfrac{81}{8}$ when $x = \tfrac{7}{4}$

We now have one point on the graph of f(x) and we know that the curve is symmetrical about this value of x. However, to locate the curve more accurately we need another point and we use f(0) as it is easy to find.

f$(0) = -4$

2 Draw a quick sketch of the graph of f$(x) = (1 - 2x)(x + 3)$, $x \in \mathbb{R}$

The coefficient of x^2 is negative, so f(x) has a greatest value.

The curve cuts the x-axis when f$(x) = 0$

When f$(x) = 0$, $(1 - 2x)(x + 3) = 0 \Rightarrow x = \tfrac{1}{2}$ or -3

The average of these values is $-\tfrac{5}{4}$, so the curve is symmetrical about $x = -\tfrac{5}{4}$

We now have enough information to draw a quick sketch, but note that this method is suitable only when the quadratic function factorises. We could, if needed, find the greatest value of f(x) using $x = -\tfrac{5}{4}$ which is the axis of symmetry.

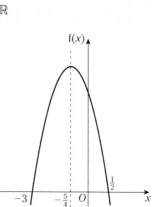

Exercise 5c

To find the greatest or least values of f(x) use factorisation when possible. Otherwise use 'completing the square'.

The domain of every function in this exercise is $x \in \mathbb{R}$

1 Find the greatest or least value of f(x) where f(x) is

(a) $x^2 - 3x + 5$

(b) $2x^2 - 4x + 5$

(c) $3 - 2x - x^2$

2 Find the range of f where f(x) is

(a) $7 + x - x^2$

(b) $x^2 - 2$

(c) $2x - x^2$

3 Sketch the graph of each of the following quadratic functions, showing the greatest or least value and the value of x at which it occurs.

(a) $x^2 - 2x + 5$

(b) $x^2 + 4x - 8$

(c) $2x^2 - 6x + 3$

(d) $4 - 7x - x^2$

(e) $x^2 - 10$

(f) $2 - 5x - 3x^2$

4 Draw a quick sketch of each of the following functions.

(a) $(x - 1)(x - 3)$

(b) $(x + 2)(x - 4)$

(c) $(2x - 1)(x - 3)$

(d) $(1 + x)(2 - x)$

(e) $x^2 - 9$

(f) $3x^2$

Cubic functions

The general form of a cubic function is

$$f(x) = ax^3 + bx^2 + cx + d, x \in \mathbb{R}$$

where a, b, c and d are constants and $a \neq 0$

Drawing the curve $y = ax^3 + bx^2 + cx + d$ for a variety of values of a, b, c and d shows that the shape of the curve is

 when $a > 0$ and   when $a < 0$

Sometimes there are no turning points and the curve looks like this

 or

Polynomial functions

The general form of a polynomial function is

$$f(x) = a_n x^n + a_{n-1} x^{n-1} + \ldots + a_2 x^2 + a_1 x + a_0, x \in \mathbb{R}$$

where $a_n, a_{n-1}, \ldots, a_0$ are rational constants, n is a positive integer and $a_n \neq 0$

Examples of polynomials are

$$f(x) = 3x^4 - 2x^3 + 5, x \in \mathbb{R}, \quad f(x) = x^5 - 2x^3 + x, x \in \mathbb{R}, \quad f(x) = x^2, x \in \mathbb{R}$$

The *order* of a polynomial is the highest power of x in the function.
So $x^4 - 7$ has order 4, and $2x - 1$ has order 1.

We have already looked at the graphs of polynomials of order 1.

e.g. $f(x) = 2x - 1$ which gives a straight line

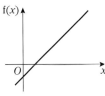

and of order 2,

e.g. $f(x) = x^2 - 4$ which gives a parabola

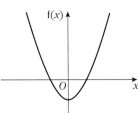

and of order 3,

e.g. $f(x) = x^3 - 2x + 1$ which gives a cubic curve.

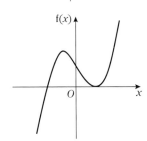

Rational functions

A rational function is one in which both numerator and denominator are polynomial.

Examples of rational functions of x are

$$\frac{1}{x}, \quad \frac{x}{x^2 - 1}, \quad \frac{3x^2 + 2x}{x - 1}$$

The simplest rational function is $f(x) = \dfrac{1}{x}$, $x \neq 0$, $x \in \mathbb{R}$, We can find various properties of $f(x)$.

1 As the value of x increases, the value of $f(x)$ gets closer to zero,

e.g. when $x = 100$, $f(x) = \frac{1}{100}$

and when $x = 1000$, $f(x) = \frac{1}{1000}$

We write this as 'when $x \to \infty$, $f(x) \to 0$'

Also as the value of x decreases, i.e. as $x \to -\infty$, the value of $f(x)$ again gets closer to zero,
i.e. when $x \to -\infty$, $f(x) \to 0$

2 $f(x)$ does not exist when $x = 0$, so this value of x must be excluded from the domain of f.

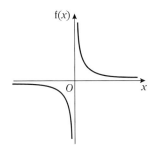

x can get as close as we like to zero, however, and can approach zero in two ways.

If $x \to 0$ from above (i.e. from positive values, 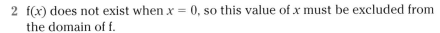) then $f(x) \to \infty$.

If $x \to 0$ from below (i.e. from negative values, 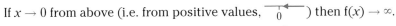) then $f(x) \to -\infty$.

As $x \to \pm\infty$, the curve gets close to the x-axis but does not cross it. Also, as $x \to 0$, the curve approaches the y-axis but again does not cross it.

The x-axis and the y-axis are called *asymptotes* to the curve.

Exercise 5d

1 Draw sketch graphs of the following functions.

(a) $x^2 + 4$

(b) $(x + 2)(x - 5)$

(c) $x^3 - 1$

(d) $(x - 1)^3$

2 Find the values of x where the curve $y = f(x)$ cuts the x-axis and sketch the curve when

(a) $f(x) = x(x - 1)(x + 1), x \in \mathbb{R}$

(b) $f(x) = x(x - 1)(x + 1)(x - 2), x \in \mathbb{R}$

(c) $f(x) = (x^2 - 1)(2 - x), x \in \mathbb{R}$

(d) $f(x) = (x^2 - 1)(4 - x^2), x \in \mathbb{R}$

INVERSE FUNCTIONS

f is the function where $f(x) = 2x$ for $x = 2, 3, 4$

The domain $\{2, 3, 4\}$ maps to the range $\{4, 6, 8\}$. This is illustrated by the arrow diagram.

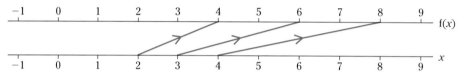

We can reverse this mapping, i.e. we can map each member of the range back to the corresponding member of the domain by halving each member of the range.

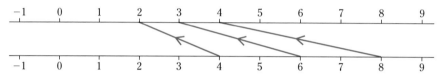

This reverse mapping can be expressed algebraically, i.e. for $x = 4, 6, 8$, $x \mapsto \frac{1}{2}x$ maps 4 to 2, 6 to 3 and 8 to 4, and it is a function in its own right. It is called the *inverse* function of f where $f(x) = 2x$

Denoting this inverse function by f^{-1} we can write $f^{-1}(x) = \frac{1}{2}x$ for $x = 4, 6, 8$

For any function f,

if there exists a function, g, that maps the output of f back to its input, i.e. $g : f(x) \mapsto x$, then this function is called the inverse of f and it is denoted by f^{-1}.

THE GRAPH OF A FUNCTION AND ITS INVERSE

The diagram shows the curve that is obtained by reflecting $y = f(x)$ in the line $y = x$. The reflection of a point $A(a, b)$ on the curve $y = f(x)$ is the point A' whose coordinates are (b, a), i.e. interchanging the x- and y-coordinates of A gives the coordinates of A'.

We can therefore obtain the equation of the reflected curve by interchanging x and y in the equation $y = f(x)$

Now the coordinates of A on $y = f(x)$ can be written as $[a, f(a)]$. Therefore the coordinates of A' on the reflected curve are $[f(a), a]$, i.e. the equation of the reflected curve is such that the output of f is mapped to the input of f.

Hence if the equation of the reflected curve can be written in the form $y = g(x)$, then g is the inverse of f, i.e. $g = f^{-1}$

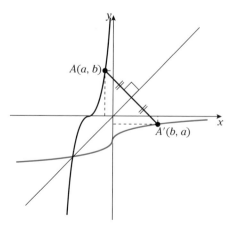

Any curve whose equation can be written in the form $y = f(x)$ can be reflected in the line $y = x$. However, this reflected curve may not have an equation that can be written in the form $y = f^{-1}(x)$

The diagram shows the curve $y = x^2$ and its reflection in the line $y = x$

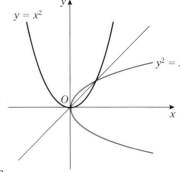

The equation of the image curve is $x = y^2 \Rightarrow y = \pm\sqrt{x}$ and $x \mapsto \pm\sqrt{x}$ is not a function.

(We can see this from the diagram as, on the reflected curve, one value of x maps to two values of y. So in this case y cannot be written as a function of x.)

A function such as $f : x \mapsto x^2$ for $x \in \mathbb{R}$, where more than one value of x maps to a value of $f(x)$, is called a many–one function.

Therefore the function $f : x \mapsto x^2$ for $x \in \mathbb{R}$ does not have an inverse, i.e.

not every function has an inverse.

If we change the definition of f to $f : x \mapsto x^2$ for $x \in \mathbb{R}^+$ then the inverse mapping is

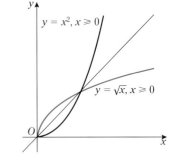

$x \mapsto \sqrt{x}$ for $x \in \mathbb{R}^+$ and this is a function, i.e. $\quad \mathbb{R}^+$ is the set of positive real numbers including zero.

$f^{-1}(x) = \sqrt{x}$ for $x \in \mathbb{R}^+$

The function $f : x \mapsto x^2$ for $x \in \mathbb{R}^+$ is such that only one value of x maps to a value of $f(x)$.

Any function where only one value of x maps to one value of y is called a one–one function.

A function f has an inverse only if f is a one–one function.

To summarise:

The inverse of a function undoes the function, i.e. it maps the output of a function back to its input.

The inverse of the function f is written f^{-1}.

Only one–one functions have an inverse.

When the curve whose equation is $y = f(x)$ is reflected in the line $y = x$, the equation of the reflected curve is $x = f(y)$

If this equation can be written in the form $y = g(x)$ then g is the inverse of f, i.e. $g(x) = f^{-1}(x)$, and the domain of g is the range of f.

Examples 5e

1 Determine whether there is an inverse of the function f given by $f(x) = 2 + \dfrac{1}{x}$, $x \neq 0$
   If f^{-1} exists, express it as a function of x.

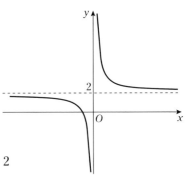

From the sketch of $f(x) = 2 + \dfrac{1}{x}$, we see that one value of $f(x)$ maps to one value of x, i.e. f is a one–one function.

Therefore the reverse mapping is a function.

The equation of the reflection of $y = 2 + \dfrac{1}{x}$ can be written as

$x = 2 + \dfrac{1}{y} \quad \Rightarrow \quad y = \dfrac{1}{x-2} \qquad$ Interchange x and y.

$\therefore \quad$ when $f(x) = 2 + \dfrac{1}{x}$, $\ f^{-1}(x) = \dfrac{1}{x-2}$, provided that $x \neq 2$

2 Find $f^{-1}(4)$ when $f(x) = 5x - 1$, $x \in \mathbb{R}$

$y = 5x - 1$

For the reflected curve $x = 5y - 1 \Rightarrow y = \frac{1}{5}(x + 1)$ and $\frac{1}{5}(x + 1)$ is a function.

i.e. $\qquad f^{-1}(x) = \frac{1}{5}(x + 1)$

$\therefore \qquad f^{-1}(4) = \frac{1}{5}(4 + 1) = 1$

Exercise 5e

1 Sketch the graphs of $f(x)$ and $f^{-1}(x)$ on the same axes.

(a) $f(x) = 3x - 1$, $x \in \mathbb{R}$

(b) $f(x) = 2 - x$, $x \in \mathbb{R}$

(c) $f(x) = \frac{1}{x} - 3$, $x \neq 0$, $x \in \mathbb{R}$

(d) $f(x) = \frac{1}{x}$, $x \neq 0$, $x \in \mathbb{R}$

2 Which of the functions given in question **1** are their own inverses?

3 Determine whether f has an inverse function, and if it does, find it when

(a) $f(x) = x + 1$, $x \in \mathbb{R}$

(b) $f(x) = x^2 + 1$, $x \in \mathbb{R}$

(c) $f(x) = x^3 + 1$, $x \in \mathbb{R}$

(d) $f(x) = x^2 - 4$, $x \geqslant 0$, $x \in \mathbb{R}$

(e) $f(x) = (x + 1)^4$, $x \geqslant -1$, $x \in \mathbb{R}$

4 The function f is given by

$f(x) = 1 - \frac{1}{x}$, $x \neq 0$, $x \in \mathbb{R}$.

Find

(a) $f^{-1}(4)$

(b) the value of x for which $f^{-1}(x) = 2$

(c) any values of x for which $f^{-1}(x) = x$

COMPOSITION OF FUNCTIONS

Two functions f and g are given by

$$f(x) = x^2, \; x \in \mathbb{R} \quad \text{and} \quad g(x) = \frac{1}{x}, \; x \neq 0, \; x \in \mathbb{R}$$

These two functions can be combined in several ways.

1 They can be added or subtracted,

i.e. $f(x) + g(x) = x^2 + \frac{1}{x}$, $x \neq 0$, $x \in \mathbb{R}$ and $f(x) - g(x) = x^2 - \frac{1}{x}$, $x \neq 0$, $x \in \mathbb{R}$

2 They can be multiplied or divided,

i.e. $f(x)g(x) = (x^2) \times \left(\frac{1}{x}\right) = x$, $x \neq 0$, $x \in \mathbb{R}$ and $\dfrac{f(x)}{g(x)} = \dfrac{x^2}{\frac{1}{x}} = x^3$, $x \neq 0$, $x \in \mathbb{R}$

3 The output of f can be made the input of g,

i.e. $x \overset{f}{\longmapsto} x^2 \overset{g}{\longmapsto} \frac{1}{x^2}$ or $g[f(x)] = g(x^2) = \frac{1}{x^2}$, $x \neq 0$, $x \in \mathbb{R}$

Therefore the function $x \mapsto \dfrac{1}{x^2}$ is obtained by taking the function g of the function f.

This is a composite function, also called a function of a function, and is written as $gf(x)$.

For $f(x) = x^2$, $x \in \mathbb{R}$ and $g(x) = 3x - 1$, $x \in \mathbb{R}$

$gf(x)$ means the function g of the function $f(x)$, i.e. $gf(x) = g(x^2) = 3x^2 - 1$, $x \in \mathbb{R}$

$fg(x)$ means the function f of the function $g(x)$, i.e. $f(3x - 1) = (3x - 1)^2$, $x \in \mathbb{R}$

This shows that the composite function fg(x) is not the same as the composite function gf(x).

For any composite function gf(x), f(x) is the range of f and this range gives the input values of g. Therefore the range of f must be included in the domain of g.

Exercise 5f

1 f, g and h are functions defined by
   $f(x) = x^2, x > 0, x \in \mathbb{R}$, $g(x) = 2x + 1, x \in \mathbb{R}$,
   $h(x) = 1 - x, x \in \mathbb{R}$
   Find as a function of x

   (a) fg   (b) fh   (c) hg

   (d) hf   (e) gf

2 $f(x) = 2x - 1, x \in \mathbb{R}$ and $g(x) = x^3, x \in \mathbb{R}$
   Find the value of

   (a) gf(3)   (b) fg(2)

   (c) fg(0)   (d) gf(0)

3 $f(x) = 2x, x \in \mathbb{R}$, $g(x) = 1 + x, x \in \mathbb{R}$, and
   $h(x) = x^2, x \in \mathbb{R}$
   Find as a function of x

   (a) hg   (b) fhg   (c) ghf

4 The function f is such that $f(x) = (2 - x)^2$, $x \in \mathbb{R}$. Find g and h as functions of x such that $gh(x) = f(x)$

5 Repeat question **4** when $f(x) = (x + 1)^4, x \in \mathbb{R}$

6 Express the function $f(x)$ as a combination of functions $g(x)$ and $h(x)$, and define $g(x)$ and $h(x)$, where $f(x)$ is

   (a) $(3x - 2)^2, x \in \mathbb{R}$

   (b) $(2x + 1)^3, x \in \mathbb{R}$

   (c) $(5x - 6)^4, x \in \mathbb{R}$

   (d) $(x - 1)(x^2 - 2), x \in \mathbb{R}$

Mixed exercise 5

1 A function f is defined by
   $f(x) = (1 - x)^2, x < 0, x \in \mathbb{R}$

   (a) Find the value of f(−3).

   (b) Sketch the curve $y = f(x)$

   (c) Find $f^{-1}(x)$ in terms of x and give the domain of f^{-1}.

2 Find the greatest or least value of each of the following functions, stating the value of x at which they occur.

   (a) $f(x) = x^2 - 3x + 5, x \in \mathbb{R}$

   (b) $f(x) = 2x^2 - 7x + 1, x \in \mathbb{R}$

   (c) $f(x) = (x - 1)(x + 5), x \in \mathbb{R}$

   State the range of f in each case.

3 $f(x) = 3x, x \in \mathbb{R}$, $g(x) = 1 - x, x \in \mathbb{R}$ and $h(x) = 3x - 1, x \in \mathbb{R}$

   Find

   (a) fg(x)

   (b) gfh(x)

   (c) $g^{-1}f^{-1}(x)$

   (d) $(gf)^{-1}(x)$

4 The function f is defined by
   $f(x) = 2x - 3, x \in \mathbb{R}$

   (a) Sketch on the same diagram the graphs of $y = f(x)$ and $y = f^{-1}(x)$

   The functions g and h are defined by

   $g : x \mapsto x^2$ for $x \in \mathbb{R}$ and $h : x \mapsto 4x$ for $x \in \mathbb{R}$

   (b) Express hgf(x) in terms of x.

   (c) Sketch the graph of $y = gf(x)$

5 The functions f and g are defined by

   $f : x \mapsto x - 1$ for $x \in \mathbb{R}$ and $g : x \mapsto \dfrac{x}{2x + 1}$ for $x \in \mathbb{R}, x \neq -\frac{1}{2}$

   (a) Express $f^{-1}(x)$ and $g^{-1}(x)$ in terms of x.

   (b) Show that the equation fg(x) = x has no solution.

6 The functions f and g are defined by

   $f : x \mapsto x - 2$ for $x \in \mathbb{R}$ and $g : x \mapsto x^2 + 4$ for $x \in \mathbb{R}$

   (a) Find the minimum value of fg(x).

   (b) Solve the equation gf(x) = 2x

7 The function f is defined by
$f : x \mapsto x^2 - 3x + 4$ for $x \in \mathbb{R}$

(a) Find the minimum value of f(x).

The function g is defined by
$g : x \mapsto x^2 - 3x + 4$ for $x \geqslant k,\ x \in \mathbb{R}$

(b) Find the smallest value of k for which $g^{-1}(x)$ exists.

(c) Sketch on the same diagram the graphs of $y = g(x)$ and $y = g^{-1}(x)$

8 The function f is defined by
$f : x \mapsto 2x - 5x^2$ for $x \in \mathbb{R}$. Find the range of f.

9 The function f is defined by
$f : x \mapsto 2x^2 - 12x + 21$ for $0 \leqslant x \leqslant k,\ x \in \mathbb{R}$

(a) Find the value of k for which the graph of $y = f(x)$ has a line of symmetry.

(b) Find, for this value of k, the range of f.

6 Inequalities and intersection of curves

After studying this chapter you should be able to

- solve linear and quadratic inequalities in one unknown
- use the relationship between points of intersection of graphs and solutions of equations (including, in simple cases, the correspondence between a line being tangent to a curve and a repeated root of an equation).

MANIPULATING INEQUALITIES

An inequality compares two unequal quantities.

For example, the two real numbers 3 and 8 for which

$$8 > 3$$

The inequality remains true, i.e. the inequality sign is unchanged, when the same term is added or subtracted on both sides, e.g.

$$8 + 2 > 3 + 2 \quad \Rightarrow \quad 10 > 5$$

and $\quad 8 - 1 > 3 - 1 \quad \Rightarrow \quad 7 > 2$

The inequality sign is unchanged also when both sides are multiplied or divided by a positive quantity, e.g.

$$8 \times 4 > 3 \times 4 \quad \Rightarrow \quad 32 > 12$$

and $\quad 8 \div 2 > 3 \div 2 \quad \Rightarrow \quad 4 > 1\frac{1}{2}$

When both sides are multiplied or divided by a *negative* quantity the inequality is no longer true. For example, if we multiply by -1, the LHS becomes -8 and the RHS becomes -3 so the correct inequality is now LHS < RHS, i.e.

$$8 \times -1 < 3 \times -1 \Rightarrow -8 < -3$$

Similarly, dividing by -2 gives $-4 < -1\frac{1}{2}$

These examples are illustrations of the following general rules.

> **Adding or subtracting a term, or multiplying or dividing both sides by a positive number, does not alter the inequality sign.**
>
> **Multiplying or dividing both sides by a *negative* number reverses the inequality sign.**
>
> **i.e. if *a*, *b* and *k* are real numbers, and *a* > *b* then**
>
> $a + k > b + k$ **for all values of *k*.**
>
> $ak > bk$ **for positive values of *k*.**
>
> $ak < bk$ **for negative values of *k*.**

SOLVING LINEAR INEQUALITIES

When an inequality contains an unknown quantity, the rules given above can be used to 'solve' it. The solution of an inequality is a range, or ranges, of values of the variable. For example, when $x - 2 > 0$, then adding 2 to each side gives $x > 2$. This gives the range of values for which the inequality is true and it is called the solution of the inequality.

When the unknown quantity appears only in linear form, we have a *linear inequality* and the solution range has only *one boundary*.

Example 6a

Find the set of values of x that satisfy the inequality $x - 5 < 2x + 1$

$x - 5 < 2x + 1 \implies x < 2x + 6$ adding 5 to each side

$\implies -x < 6$ subtracting $2x$ from each side

$\implies x > -6$ multiplying both sides by -1

So the set of values of x satisfying the given inequality is $x > -6$

Exercise 6a

Solve the following inequalities.

1 $x - 4 < 3 - x$

2 $7 - 3x < 13$

3 $1 - 7x > x + 3$

4 $x + 3 < 3x - 5$

5 $x > 5x - 2$

6 $2(3x - 5) > 6$

7 $x < 4x + 9$

8 $2x - 1 < x - 4$

9 $3(3 - 2x) < 2(3 + x)$

SOLVING QUADRATIC INEQUALITIES

A quadratic inequality is one in which the variable appears to the power 2, e.g. $x^2 - 3 > 2x$

The solution is a range or ranges of values of the variable with *two boundaries*.

If the terms in the inequality can be collected and factorised, a graphical solution is easy to find.

Examples 6b

1 Find the range(s) of values of x that satisfy the inequality $x^2 - 3 > 2x$

$$x^2 - 3 > 2x \implies x^2 - 2x - 3 > 0$$

$$\implies (x - 3)(x + 1) > 0$$

or $f(x) > 0$ where $f(x) = (x - 3)(x + 1)$

A sketch of the graph of $f(x)$ shows that $f(x) > 0$ where the graph is above the x-axis. The values of x corresponding to these portions of the graph satisfy $f(x) > 0$. The points where $f(x) = 0$, i.e. where $x = 3$ and -1, are not part of this solution and this is indicated on the sketch by open circles.

From the graph we see that the ranges of values of x which satisfy the given inequality are $x < -1$ and $x > 3$

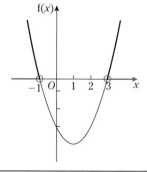

Investigating the nature of the roots of a quadratic equation can result in a quadratic inequality.

Examples 6b cont.

2 Find the range(s) of values of k for which the roots of the equation $kx^2 + kx - 2 = 0$ are real.

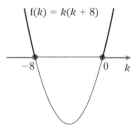

$kx^2 + kx - 2 = 0$

For real roots '$b^2 - 4ac$' $\geqslant 0$

i.e. $\qquad k^2 - 4(k)(-2) \geqslant 0$

$\Rightarrow \qquad k(k + 8) \geqslant 0$

Therefore the equation $kx^2 + kx - 2 = 0$ has real roots if the value of k lies in either of the ranges $k \leqslant -8$ or $k \geqslant 0$

The sketch shows solid circles where $k = -8$ and $k = 0$ because these values are part of the solution.

Exercise 6b

Find the ranges of values of x that satisfy the following inequalities.

1 $(x - 2)(x - 1) > 0$

2 $(x + 3)(x - 5) \geqslant 0$

3 $(x - 2)(x + 4) < 0$

4 $(2x - 1)(x + 1) \geqslant 0$

5 $x^2 - 4x > 3$

6 $4x^2 < 1$

7 $(2 - x)(x + 4) \geqslant 0$

8 $5x^2 > 3x + 2$

9 $(3 - 2x)(x + 5) \leqslant 0$

10 $(x - 1)^2 > 9$

11 $(x + 1)(x + 2) \leqslant 4$

12 $(1 - x)(4 - x) > x + 11$

13 Find the values of p for which the given equation has real roots.
  (a) $x^2 + (p + 3)x + 4p = 0$
  (b) $x^2 + 3x + 1 = px$

14 Find the range of values of a for which the equation $x^2 - ax + (a + 3) = 0$ has no real roots.

15 What is the set of values of p for which $p(x^2 + 2) < 2x^2 + 6x + 1$ for all real values of x?

16 The function f is defined by $f : x \mapsto x^2 + kx + 9$ for $x \in \mathbb{R}$

Find the range of values of k for which the range of f is $f(x) \geqslant 0$

INTERSECTION OF A LINE AND A PARABOLA

A straight line can intersect a parabola at two separate points or it can just touch the parabola at one point or it might not intersect at any point.

Solving the equation of the parabola and the equation of the line simultaneously finds the points of intersection.

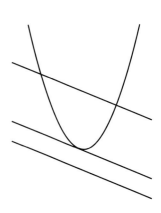

Solving $y = ax^2 + bx + c$ and $y = mx + c$ simultaneously gives the quadratic equation $ax^2 + bx + c = mx + c$

When this equation has two separate roots, the line cuts the parabola at two separate points.

When this equation has a repeated root (only one solution), the line touches the parabola at one point, i.e. the line is a tangent to the curve.

When this equation has no real roots, the line does not cut the parabola at any point.

The same argument can be used for the intersection of any curve and a straight line. When solving the equations of the line and curve simultaneously gives a quadratic equation, the nature of the roots of that equation tells you whether the line cuts the curve, is a tangent to the curve or does not cut the curve.

Examples 6c

1 Determine the set of values of c for which the line $y = 2x + c$ does not intersect the curve $y = 2x^2 - 3x + 2$

Solving $y = 2x + c$ and $y = 2x^2 - 3x + 2$ simultaneously gives

$$2x^2 - 3x + 2 = 2x + c$$
$$\Rightarrow \quad 2x^2 - 5x + 2 - c = 0$$

The line does not intersect the curve when this equation has no real roots, i.e. when

$$\text{'}b^2 - 4ac < 0\text{'} \quad \Rightarrow \quad 25 - 8(2 - c) < 0$$
$$\Rightarrow \quad c < -\frac{9}{8}$$

2 Determine the values of m for which the line $y = mx - 5$ is a tangent to the curve $y = x^2 - 1$

Solving $y = x^2 - 1$ and $y = mx - 5$ simultaneously gives

$$x^2 - 1 = mx - 5$$
$$\Rightarrow \quad x^2 - mx + 4 = 0$$

The line is a tangent to the curve when this equation has a repeated root (i.e. one solution),

$$\Rightarrow \quad m^2 - 16 = 0$$
$$\therefore \quad m = \pm 4$$

Exercise 6c

1 Find the range of values of k for which the line intersects the curve at two separate points when the equation of the curve and the equation of the line are

(a) $y = x^2 + k, y = 2x - 1$

(b) $y = kx^2 - 6, y = x - 5$

2 Find the range of values of k for which the line does not intersect the curve when the equation of the curve and the equation of the line are

(a) $y = x^2 + kx + 5, y = 5x$

(b) $y = kx^2 - 2x + k, y = x - 3$

3 Find the values of k for which the line is a tangent to the curve when the equation of the curve and the equation of the line are

(a) $y = x^2 - 4x + 4, y = 2x + k$

(b) $y = kx^2 + x + 1, y = 5 - x$

4 Show that there are no values of k for which the line $y = 3 - 2x$ can intersect the curve $y = 2x^2 + kx + 3$

5 Functions f and g are defined by

f: $x \mapsto x^2 - 4$ for $x \in \mathbb{R}$ and
g: $x \mapsto x - 2$ for $x \in \mathbb{R}$

Find the range of values of k for which the curve $y = fg(x)$ and the line $y = x - k$ intersect at two separate points.

6 Show that the line $y = 2x - 1$ is a tangent to the curve $y = x^2$

7 Show that the line $y = 3 - x$ is a tangent to the curve $y = \dfrac{1}{x - 1}$

8 A function f is defined by $f : x \mapsto \dfrac{1}{x - 2}$ for $x \in \mathbb{R}$ and $x > 2$

Show that the line $y = 1 - \frac{1}{2}x$ does not intersect the curve $y = f(x)$

Mixed exercise 6

Solve each of the inequalities given in questions **1** to **10**.

1 $2x + 1 < 4 - x$

2 $x - 5 > 1 - 3x$

3 $6x - 5 > 1 + 2x$

4 $(x - 3)(x + 2) > 0$

5 $(2x - 3)(3x + 2) < 0$

6 $x^2 - 3 < 10$

7 $(x - 3)^2 > 2$

8 $(3 - x)(2 - x) < 20$

9 $x(4x + 3) > 2x - 1$

10 $(x - 6)(x + 1) > 2x - 12$

11 For what values of k does the equation $4x^2 + 8x - 8 = k(4x - 3)$ have real roots?

12 (a) Show that, for all values of k, the line $y = 3x - k$ intersects the curve $y = \dfrac{1}{x - 1}$ at two points.

(b) Find the values of k for which the line $y = kx + 3$ is a tangent to the curve $y = \dfrac{1}{x - 1}$

7 Differentiation

After studying this chapter you should be able to

- understand the idea of the gradient of a curve, and use the notations $f'(x)$ and $\dfrac{dy}{dx}$
- use the derivative of x^n for any rational n, together with constant multiples, sums, differences of functions, and of composite functions using the chain rule
- apply differentiation to gradients and normals, increasing and decreasing functions and rates of change (including connected rates of change).

CHORDS, TANGENTS, NORMALS AND GRADIENTS

A and B are any two points on any curve.

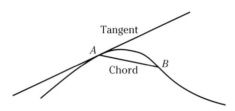
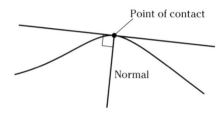

The line joining A and B is called a chord.

The line that touches the curve at A is called the tangent at A.

The word *touch* has a precise mathematical meaning. A line that meets a curve at a point and does not cross to the other side of the curve at that point, *touches* the curve at the *point of contact*.

The line perpendicular to the tangent at A is called the normal at A.

Gradient defines the direction of a line (lines can be straight or curved).

The gradient of a straight line is constant.

Moving from B to A along the curve in the diagram, the direction is changing all the time. If at A we continue to move, but without any further change in direction, we go along the straight line AT, i.e. along the tangent to the curve at A, so

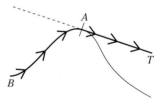

the gradient of the curve at A is the same as the gradient of the tangent to the curve at A.

When the line or curve is drawn on a pair of x- and y-axes, the gradient is the rate at which y increases with respect to x.

For a straight line this is found by taking the coordinates of two points and working out $\dfrac{\text{increase in } y}{\text{increase in } x}$.

For a curve, start with two points A and B on the curve. When A and B are fairly close, the gradient of the chord AB gives an approximate value for the gradient of the tangent at A. As B gets closer to A, the chord AB gets closer to the tangent at A so the approximation becomes more accurate.

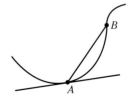

Hence   **as $B \to A$**

the gradient of chord $AB \to$ the gradient of the tangent at A

or

$\displaystyle\lim_{\text{as } B \to A}$ (gradient of chord AB) = gradient of tangent at A

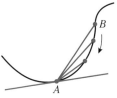

THE GRADIENT AT ANY POINT ON THE CURVE $y = x^2$

$A(x, y)$ is any point on the curve $y = x^2$ and B is a point close to A. The x-coordinate of B is $x + \delta x$

δx means a small increase in the value of x.

For any point on the curve, $y = x^2$

So, at B, the y-coordinate is $(x + \delta x)^2 = x^2 + 2x\delta x + (\delta x)^2$

The gradient of chord AB is given by $\dfrac{\text{increase in } y}{\text{increase in } x}$,

which is $\dfrac{(x + \delta x)^2 - x^2}{(x + \delta x) - x} = \dfrac{2x\delta x + (\delta x)^2}{\delta x}$

$$= 2x + \delta x$$

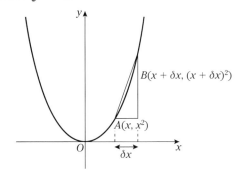

Now as $B \to A$, $\delta x \to 0$, therefore

the gradient of the curve at $A = \displaystyle\lim_{\text{as } B \to A}$ (gradient of chord AB)

$$= \lim_{\text{as } \delta x \to 0} (2x + \delta x)$$

$$= 2x$$

This process is called differentiation from first principles. You will not be asked to use it in the examination.

This result can now be used to find the gradient at any particular point on the curve with equation $y = x^2$

e.g. at the point where $x = 3$, the gradient is $2(3) = 6$ and at the point $(4, 16)$, the gradient is $2(4) = 8$

The process of finding a general expression for the gradient of a curve at any point is known as differentiation. The general expression for the gradient of a curve $y = f(x)$ is itself a function so it is called the *gradient function*. For the curve $y = x^2$, for example, the gradient function is $2x$.

Because the gradient function is derived from the given function, it is usually called the *derived function* or the *derivative*.

Notation

The notation for the derivative of a function $f(x)$ is $f'(x)$

So when $f(x) = x^2$ we write $f'(x) = 2x$

The notation for the gradient of a curve $y = f(x)$ is $\dfrac{dy}{dx} = f'(x)$

So for $y = x^2$, we write $\dfrac{dy}{dx} = 2x$

The symbol d is not a factor as it has no meaning on its own.

The complete symbol $\dfrac{d}{dx}$ means 'the derivative with respect to x of'

So $\dfrac{dy}{dx}$ means 'the derivative with respect to x of y'

and $\dfrac{d}{dx}(x^2 - 3x)$ means 'the derivative with respect to x of $x^2 - 3x$'.

The method used to find the gradient function of $y = x^2$ can be used to find the gradient function of any curve, i.e.
$\dfrac{dy}{dx} = \displaystyle\lim_{\delta x \to 0}\left(\dfrac{f(x + \delta x) - f(x)}{\delta x}\right)$. However, you will not be asked to use this in the examination.

DIFFERENTIATING x^n WITH RESPECT TO x

For any rational number n, differentiating $y = x^n$ gives $\dfrac{dy}{dx} = nx^{n-1}$

This rule can be verified for a particular value of n by following the same process used to find the derivative of x^2.

Therefore, for example, when $y = x^4$, $\dfrac{dy}{dx} = 4x^3$

Example 7a

Differentiate with respect to x

(a) $x^{-\frac{1}{3}}$ (b) $\sqrt[4]{(x^3)}$

(a) Use $\dfrac{d}{dx}(x^n) = nx^{n-1}$, where $n = -\dfrac{1}{3}$

$$\frac{dy}{dx} = -\frac{1}{3}x^{-\frac{1}{3}-1} = -\frac{1}{3}x^{-\frac{4}{3}} = -\frac{1}{3x^{\frac{4}{3}}}$$

(b) $\sqrt[4]{(x^3)}$ can be written $x^{\frac{3}{4}}$, i.e. $n = \dfrac{3}{4}$

$$\frac{d}{dx}\left(x^{\frac{3}{4}}\right) = \frac{3}{4}x^{\frac{3}{4}-1} = \frac{3}{4}x^{-\frac{1}{4}} \text{ or } \frac{3}{4\sqrt[4]{x}}$$

Exercise 7a

Differentiate with respect to x

1 x^5

2 x^{-3}

3 $x^{\frac{4}{3}}$

4 $\dfrac{1}{x}$

5 x^{10}

6 $\dfrac{1}{x^2}$

7 $\sqrt{x^3}$

8 $x^{-\frac{1}{2}}$

9 $\dfrac{1}{x^4}$

10 $x^{\frac{1}{3}}$

11 $x^{-\frac{1}{4}}$

12 x

13 $\sqrt{x^7}$

14 $\dfrac{1}{x^7}$

15 $x^{\frac{1}{7}}$

16 $\sqrt{(x^2)^3}$

DIFFERENTIATING CONSTANTS AND MULTIPLES OF x

Any line with equation $y = c$, where c is a constant, is a horizontal straight line whose gradient is zero,

$\therefore$ if $y = c$ then $\dfrac{dy}{dx} = 0$

Any line with equation $y = kx$, where k is a constant, is a sloping line with gradient k,

$\therefore$ if $y = kx$ then $\dfrac{dy}{dx} = k$

When y is a constant multiple of a function of x, i.e. $y = af(x)$ then $\dfrac{dy}{dx} = af'(x)$

e.g.   if $y = 3x^5$, $\dfrac{dy}{dx} = 3 \times 5x^4 = 15x^4$

and   if $y = 4x^{-2}$, $\dfrac{dy}{dx} = 4 \times -2x^{-3} = -8x^{-3}$

In general, if a is a constant

$$\frac{d}{dx} ax^n = anx^{n-1}$$

A function of x that contains the sum or difference of a number of separate terms can be differentiated term by term, applying the basic rule to each in turn.

For example, if $y = x^4 + \dfrac{1}{x} - 6x$

then $\dfrac{dy}{dx} = \dfrac{d}{dx}(x^4) + \dfrac{d}{dx}(x^{-1}) - \dfrac{d}{dx}(6x) = 4x^3 - \dfrac{1}{x^2} - 6$

Exercise 7b

Differentiate each of the following functions with respect to x.

1 $x^3 - x^2 + 5x - 6$

2 $3x^2 + 7 - \dfrac{4}{x}$

3 $\sqrt{x} + \dfrac{1}{\sqrt{x}}$

4 $2x^4 - 4x^2$

5 $x^3 - 2x^2 - 8x$

6 $x^2 + 5\sqrt{x}$

7 $x^{-\frac{3}{4}} - x^{\frac{3}{4}} + x$

8 $3x^3 - 4x^2 + 9x - 10$

9 $x^{\frac{3}{2}} - x^{\frac{1}{2}} + x^{-\frac{1}{2}}$

10 $\sqrt{x} + \sqrt{x^3}$

11 $\dfrac{1}{x^2} - \dfrac{1}{x^3}$

12 $\dfrac{1}{\sqrt{x}} - \dfrac{2}{x}$

13 $x^{-\frac{1}{2}} + 3x^{\frac{3}{2}}$

14 $x^{\frac{1}{4}} - x^{\frac{1}{5}}$

15 $\dfrac{4}{x^3} + \dfrac{x^3}{4}$

16 $\dfrac{4}{x} + \dfrac{5}{x^2} - \dfrac{6}{x^3}$

17 $3\sqrt{x} - 3x$

18 $x - 2x^{-1} - 3x^{-3}$

19 $x\sqrt{x} - x^2\sqrt{x}$

20 $\dfrac{\sqrt{x}}{x^2} + \dfrac{x^2}{\sqrt{x}}$

In questions 21 to 28, multiply out the brackets before you differentiate.

21 $y = (x + 1)^2$

22 $y = x^{-2}(2 - x)$

23 $y = (3x - 4)(x + 5)$

24 $y = (4 - x)^2$

25 $y = \left(\dfrac{1}{x}\right)(x^2 + 1)$

26 $y = 2x(3x^2 - 4)$

27 $y = (x + 2)(x - 2)$

28 $y = x^2(x - 1)$

GRADIENTS OF TANGENTS AND NORMALS

When the equation of a curve is known, and the gradient function can be found, then the gradient, m say, at a particular point A on that curve can be calculated. This is also the gradient of the tangent to the curve at A.

The normal at A is perpendicular to the tangent at A, therefore its gradient is $-\dfrac{1}{m}$

Examples 7c

1 The equation of a curve is $s = 6 - 3t - 4t^2 - t^3$. Find the gradient of the tangent and of the normal to the curve at the point $(-2, 4)$.

$s = 6 - 3t - 4t^2 - t^3 \Rightarrow \dfrac{ds}{dt} = 0 - 3 - 8t - 3t^2$

At the point $(-2, 4)$, $\dfrac{ds}{dt} = -3 - 8(-2) - 3(-2)^2 = 1$

Therefore the gradient of the tangent at $(-2, 4)$ is 1 and the gradient of the normal is $\dfrac{-1}{1}$, i.e. -1.

2 Find the coordinates of the points on the curve $y = 2x^3 - 3x^2 - 8x + 7$ where the gradient is 4.

$y = 2x^3 - 3x^2 - 8x + 7 \Rightarrow \dfrac{dy}{dx} = 6x^2 - 6x - 8$

When the gradient is 4, then $\dfrac{dy}{dx} = 4$

i.e. $6x^2 - 6x - 8 = 4 \Rightarrow 6x^2 - 6x - 12 = 0 \Rightarrow x^2 - x - 2 = 0$

$\therefore (x - 2)(x + 1) = 0 \Rightarrow x = 2$ or -1

When $x = 2$, $y = 16 - 12 - 16 + 7 = -5$

when $x = -1$, $y = -2 - 3 + 8 + 7 = 10$

Therefore the gradient is 4 at the points $(2, -5)$ and $(-1, 10)$.

Exercise 7c

Find the gradient of the tangent and the gradient of the normal at the given point on the given curve.

1 $y = x^2 + 4$ where $x = 1$

2 $y = \dfrac{3}{x}$ where $x = -3$

3 $y = \sqrt{x}$ where $x = 4$

4 $y = 2x^3$ where $x = -1$

5 $y = 2 - \dfrac{1}{x}$ where $x = 1$

6 $y = (x + 3)(x - 4)$ where $x = 3$

7 $y = x^3 - x$ where $x = 2$

8 $y = x + x^2$ where $x = -2$

9 $y = x^2 - \dfrac{2}{x}$ where $x = 1$

10 $y = \sqrt{x} + \dfrac{1}{\sqrt{x}}$ where $x = 9$

11 $y = \dfrac{x^2 - 4}{x}$ where $x = -2$

Find the coordinates of the point(s) on the given curve where the gradient has the value specified.

12 $y = 3 - \dfrac{2}{x}; \ \dfrac{1}{2}$

13 $y = x^2 - x^3; \ -1$

14 $s = t^3 - 12t + 9; \ 15$

15 $s = t + \dfrac{1}{t}; \ 0$

16 $s = (t + 3)(t - 5); \ 0$

17 $y = \dfrac{1}{x^2}; \ \dfrac{1}{4}$

18 $y = (2x - 5)(x + 1); \ -3$

19 $y = x^3 - 3x; \ 0$

INCREASING AND DECREASING FUNCTIONS

The gradient of a curve at any point, $\dfrac{dy}{dx}$, measures the rate at which y is increasing with respect to x.

When $y = f(x)$, $\dfrac{d}{dx} f(x)$, i.e $f'(x)$, gives the rate at which $f(x)$ is increasing with respect to x.

$\therefore$ $f'(x) > 0$ **when f is increasing and** $f'(x) < 0$ **when f is decreasing**

So to determine whether a function is increasing or decreasing, we need to determine whether $f'(x)$ is positive or negative.

For example, given that $f : x \mapsto x^2$ for $x \in \mathbb{R}$

$f'(x) = 2x \Rightarrow 2x > 0$ when $x > 0$ and $2x < 0$ when $x < 0$

Therefore f is increasing when $x > 0$ and decreasing when $x < 0$

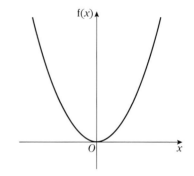

The graph of $f(x) = x^2$ confirms that the function f is increasing when $x > 0$ and decreasing when $x < 0$

Examples 7d

1 A function f is defined by $f : x \mapsto x^3 + 2x, x \in \mathbb{R}$. Show that f is an increasing function.

$$f(x) = x^3 + 2x$$
so $f'(x) = 3x^2 + 2$

x^2 is positive for all values of x,

$\therefore f'(x) > 0$ for $x \in \mathbb{R}$, so f is an increasing function.

2 Find the range of values of x for which the function f defined by $f : x \mapsto 2x^3 - 9x^2 + 12x + 4, x \in \mathbb{R}$, is a decreasing function.

$$f'(x) = 6x^2 - 18x + 12$$
$$= 6(x^2 - 3x + 2)$$
$$= 6(x - 1)(x - 2)$$

A sketch of the graph of $f'(x)$ shows that $f'(x) < 0$ for $1 < x < 2$

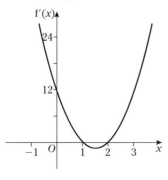

Therefore f is a decreasing function for $1 < x < 2$

Exercise 7d

1 A function f is defined by $f : x \mapsto \dfrac{1}{x}$,
   for $x \in \mathbb{R}, x \neq 0$
   Show that f is a decreasing function.

2 The equation of a curve is $y = x^5 + 4x$
   Show that the value of y increases when x increases.

3 A function f is defined by $f : x \mapsto -\dfrac{1}{x}, x > 0$
   Show that f is an increasing function.

4 Find the range of values of x for which $f : x \mapsto 3 + 4x - x^2$ is a decreasing function.

5 Find the range of values of x for which the function f defined by $f : x \mapsto 2x^3 - 3x^2 + 12x$ is an increasing function.

THE CHAIN RULE

To differentiate $(2x - 1)^3$, we can expand the bracket and then differentiate term by term. However, this takes time and, for higher powers of x, it takes a very long time.

$(2x - 1)^3$ is a composite function. Using $f(x) = 2x - 1$ we can write $y = (2x - 1)^3$ as $y = (f(x))^3$. When we make the substitution $u = f(x)$, then $y = (2x - 1)^3$ can be expressed in two parts, i.e. $u = 2x - 1$ and $y = u^3$

Then to find $\dfrac{dy}{dx}$ we can use the rule

$$\frac{dy}{dx} = \frac{dy}{du} \times \frac{du}{dx}$$

This is called the chain rule.

So, when $y = (2x - 1)^3 \quad \Rightarrow \quad u = 2x - 1$ and $y = u^3$

$$\frac{dy}{dx} = 3u^2 \times 2 = 6(2x - 1)^2$$

Proof of the chain rule:
A small increase of δx in the value of x causes a corresponding small increase of δu in the value of u.
Then if $\delta x \to 0$, it follows that $\delta u \to 0$

Hence $\qquad \dfrac{dy}{dx} = \lim_{\delta x \to 0} \left(\dfrac{\delta y}{\delta x} \right) = \lim_{\delta x \to 0} \left(\dfrac{\delta y}{\delta u} \times \dfrac{\delta u}{\delta x} \right)$

$\Rightarrow \qquad \dfrac{dy}{dx} = \left(\lim_{\delta u \to 0} \dfrac{\delta y}{\delta u} \right) \times \left(\lim_{\delta x \to 0} \dfrac{\delta u}{\delta x} \right)$

i.e. $\qquad \dfrac{dy}{dx} = \dfrac{dy}{du} \times \dfrac{du}{dx}$

Examples 7e

1 Find $\dfrac{dy}{dx}$ if $y = (2x - 4)^4$

Using $u = 2x - 4$ then $y = u^4$

$\dfrac{dy}{dx} = \dfrac{dy}{du} \times \dfrac{du}{dx}$ gives

$$\frac{dy}{dx} = (4u^3)(2) = 8u^3$$

But $u = 2x - 4$

$\therefore \qquad \dfrac{dy}{dx} = 8(2x - 4)^3$

This example is a particular case of the equation $y = (ax + b)^n$. Using the chain rule shows that

$$\textbf{when } y = (ax + b)^n \textbf{ then } \frac{dy}{dx} = an(ax + b)^{n-1}$$

This fact is quotable, e.g. $\frac{d}{dx}(3x - 2)^5 = 3(5)(3x - 2)^4$

Examples 7e cont.

2 Given $y = (x^3 + 1)^4$ find $\dfrac{dy}{dx}$

When $u = x^3 + 1$ then $y = u^4$

Using $\dfrac{dy}{dx} = \dfrac{dy}{du} \times \dfrac{du}{dx}$ gives

$$\frac{dy}{dx} = (4u^3)(3x^2) = 12x^2 u^3$$

Replacing u by $x^3 + 1$ we have

$$\frac{dy}{dx} = 12x^2(x^3 + 1)^3$$

3 Differentiate with respect to x the function $\dfrac{1}{(1 - x^2)^5}$

$$y = (1 - x^2)^{-5} \quad \Rightarrow \quad y = u^{-5} \text{ where } u = 1 - x^2$$

$$\frac{dy}{dx} = \frac{dy}{du} \times \frac{du}{dx}$$

gives $\dfrac{d}{dx}\left[\dfrac{1}{(1 - x^2)^5} \right] = (-5u^{-6})(-2x) = 10x(u^{-6})$

$$= \frac{10x}{(1 - x^2)^6}$$

You will find that with practice the necessary substitution can be done mentally and the answer written down directly, e.g. to differentiate $(x^3 - x)^{\frac{3}{2}}$, mentally use the substitutions $u = x^3 - x$ and $y = u^{\frac{3}{2}}$ giving

$$\frac{d}{dx}(x^3 - x)^{\frac{3}{2}} = \left[\tfrac{3}{2}(x^3 - x)^{\frac{1}{2}} \right](3x^2 - 1)$$

Exercise 7e

Use a substitution to differentiate each expression with respect to x.

1 $(3x + 1)^2$

2 $(3 - x)^4$

3 $(4x - 5)^5$

4 $(x^2 + 1)^3$

5 $(2 + 3x)^7$

6 $(2 - 6x)^3$

7 $(2x^4 - 5)^{\frac{1}{2}}$

8 $(x^2 + 3)^{-1}$

9 $\sqrt{3x^3 - 4}$

10 $\dfrac{1}{1 + 3x}$

11 $\dfrac{3}{4 - x^2}$

12 $\dfrac{1}{x + 1} + x$

Differentiate each function directly.

13 $(4 - 2x)^5$

14 $(x^2 + 3)^2$

15 $(3x - 4)^7$

16 $(x^2 + 4)^2$

17 $(1 - 2x^2)^3$

18 $(2 - x^3)^4$

19 $(2 + x^2)^{\frac{3}{4}}$

20 $(x^2 - x)^3$

21 $(2 - 3x^2)^{-1}$

22 $(4 - x^2)^{-2}$

23 $\sqrt{(x^5 - 3)}$

24 $\sqrt{(x - 1)} - x$

CONNECTED RATES OF CHANGE

The chain rule is useful for solving problems where, for example, we know that $y = f(x)$ and we know the rate of change of y with respect to u and want to find the rate of change of x with respect to u.

You may also need to use the fact that $\dfrac{dy}{dx} = \dfrac{1}{\dfrac{dx}{dy}}$

Example 7f

The volume, $V \text{cm}^3$, of a sphere of radius r cm is given by $V = \frac{4}{3}\pi r^3$

V is increasing at the rate of 1.5 cm^3 per second. Find the rate of increase of r with respect to time when $r = 2$

$\dfrac{dV}{dt} = 1.5$ and $V = \frac{4}{3}\pi r^3 \Rightarrow \dfrac{dV}{dr} = 4\pi r^2$

Using $\dfrac{dV}{dr} = \dfrac{dV}{dt} \times \dfrac{dt}{dr}$ gives $4\pi r^2 = 1.5 \times \dfrac{dt}{dr}$

$\Rightarrow \dfrac{dt}{dr} = \dfrac{4\pi r^2}{1.5}$

When $r = 2$, $\dfrac{dt}{dr} = \dfrac{16\pi}{1.5}$

$\therefore$ using $\dfrac{dr}{dt} = \dfrac{1}{\dfrac{dt}{dr}}$, $\dfrac{dr}{dt} = \dfrac{1.5}{16\pi} = 0.0298$ (3 s.f.)

The radius is increasing at the rate of 0.0298 cm per second.

Exercise 7f

1 The equation of a curve is $y = x^2 - 5x$.
A point P is moving along the curve so that the x-coordinate is increasing at the constant rate of 0.2 units per second.

Find the rate at which the y-coordinate is increasing when $x = 4$

2 The equation of a curve is $y = 4 - \dfrac{1}{x}$
A point P is moving along the curve so that the y-coordinate is increasing at the constant rate of 0.01 units per second.

Find the rate at which the x-coordinate is increasing when $x = 1$

3 The equation of a curve is $y = \dfrac{1}{2 - x}$
A point P is moving along the curve so that the y-coordinate is increasing at the constant rate of 0.5 units per second.

Find the rate at which the x-coordinate is increasing when $x = 1$

4 The volume, $V \text{cm}^3$, of a cube of edge x cm is increasing at the constant rate of 3 cm^3 per second. Find the rate at which x is increasing when $x = 10$

5 The volume, $V \text{cm}^3$, of a sphere of radius x cm is given by $V = \frac{4}{3}\pi x^3$. The volume is increasing at the constant rate of 0.2 cm^3 per second. Find the rate of increase of the radius when the radius is 5 cm.

6 The volume, $V \text{cm}^3$, of a sphere of radius x cm is given by $V = \frac{4}{3}\pi x^3$. The radius is increasing at the constant rate of 0.01 cm per second. Find the rate of increase of the volume when the radius is 0.8 cm.

7 A cuboid has a square base of side x cm. The height of the cuboid is 10 cm. Find the rate at which the volume, $V \text{cm}^3$, is increasing when x is increasing at the constant rate of 0.04 cm per second and $x = 10$

Mixed exercise 7

1 Differentiate $3x^2 + x$ with respect to x.

2 Find the derivative of

(a) $x^{-3} - x^3 + 7$

(b) $x^{\frac{1}{2}} - x^{-\frac{1}{2}}$

(c) $\dfrac{1}{x^2} + \dfrac{2}{x^3}$

(d) $(4x - 2)^4$

(e) $(3x - 1)^{-1}$

(f) $\dfrac{1}{x^2 + 2}$

3 Differentiate with respect to x.

(a) $y = (3x^3 - 2)^4$

(b) $y = \sqrt{x} - \dfrac{1}{x} + \dfrac{1}{x^3}$

(c) $\dfrac{1}{x} - \dfrac{1}{\sqrt{x}}$

4 Find the gradient of the curve
$y = 2x^3 - 3x^2 + 5x - 1$ at the point

(a) $(0, -1)$

(b) $(1, 3)$

(c) $(-1, -11)$

5 Find the gradient of the given curve at the given point.

(a) $y = x^2 + x - 9$; $x = 2$

(b) $s = t(t - 4)$; $t = 5$

6 The equation of a curve is $y = (x - 3)(x + 4)$.
Find the gradient of the curve

(a) at the point where the curve crosses the y-axis

(b) at each of the points where the curve crosses the x-axis.

7 The equation of a curve is $y = 2x^2 - 3x - 2$.
Find

(a) the gradient at the point where $x = 0$

(b) the coordinates of the points where the curve crosses the x-axis

(c) the gradient at each of the points found in part (b).

8 Find the coordinates of the point(s) on the curve $y = 3x^3 - x + 8$ at which the gradient is

(a) 8   (b) 0

9 Find $\dfrac{\mathrm{d}y}{\mathrm{d}x}$ when

(a) $y = x^4 - x^2$

(b) $y = (3x + 4)^2$

(c) $y = \dfrac{3}{\sqrt{x} - 1}$

10 Find the gradient of the tangent at the point where $x = 2$ on the curve $y = (2 - \sqrt{x})^2$

11 Find the coordinates of the point on the curve $y = x^2$ where the gradient of the normal is $\frac{1}{4}$.

12 The equation of a curve is $y = 4x^2 + 5x$. Find the gradient of the normal at each of the points where the curve crosses the x-axis.

13 Find the coordinates of the points on the curve $y = x^3 - 6x^2 + 12x + 2$ at which the tangent is parallel to the line $y = 3x$

14 The curve $y = (x - 2)(x - 3)(x - 4)$ cuts the x-axis at the points $P(2, 0)$, $Q(3, 0)$ and $R(4, 0)$. Prove that the tangents at P and R are parallel and find the gradient of the normal at Q.

15 The equation of a curve is $y = x + (2x + 1)^4$.

(a) Express $\dfrac{\mathrm{d}y}{\mathrm{d}x}$ in terms of x.

(b) Find the coordinates of the point on the curve at which the gradient is 9.

16 Given that $\mathrm{f} : x \mapsto (ax + b)^3, x \in \mathbb{R}$, and that when $x = 2$, $\mathrm{f}(x) = 1$ and $\mathrm{f}'(x) = 6$, find the values of a and b.

17 Given that $\mathrm{f} : x \mapsto x^{-1}, x \neq 0, x \in \mathbb{R}$ and $\mathrm{g} : x \mapsto x^2 + 2, x \in \mathbb{R}$

(a) find an expression for $\mathrm{fg}(x)$.

The equation of a curve is $y = \mathrm{fg}(x)$.

(b) Find an expression for $\dfrac{\mathrm{d}y}{\mathrm{d}x}$ in terms of x.

18 The equation of a curve is $y = x^3 - 5$

    (a)   Explain why $x^3 - 5$ is an increasing function.

    (b)   Given that x is increasing at the constant rate of 0.1 units per second, find the rate at which y is increasing when $x = 2$

19 The equation of a curve is $y = 2x - \dfrac{1}{(3x - 2)}$

    (a)   Show that $f(x) = 2x - \dfrac{1}{(3x - 2)}$ is an increasing function.

    (b)   Find the rate at which y is increasing when x is increasing at the constant rate of 0.01 units per second and $x = 2$

8 Tangents, normals and stationary values

After studying this chapter you should be able to

- use the notations $f''(x)$ and $\dfrac{d^2y}{dx^2}$
- apply differentiation to tangents and normals
- locate stationary points, and use information about stationary points in sketching graphs
- distinguish between maximum points and minimum points.

THE EQUATIONS OF TANGENTS AND NORMALS

We have seen how to find the gradient of a tangent at a particular point, A, on a curve. We also know that the tangent passes through the point A. Therefore the tangent is a line passing through a known point and having a known gradient and its equation can be found using $y - y_1 = m(x - x_1)$

The equation of a normal can be found in the same way.

Examples 8a

1 Find the equation of the normal to the curve $y = \dfrac{4}{x}$ at the point where $x = 1$

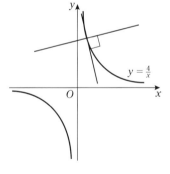

$y = \dfrac{4}{x} \quad \Rightarrow \quad \dfrac{dy}{dx} = -\dfrac{4}{x^2}$

When $x = 1$, $y = 4$ and $\dfrac{dy}{dx} = -4$

The gradient of the tangent at $(1, 4)$ is -4, therefore the gradient of the normal at $(1, 4)$ is $-\dfrac{1}{-4}$, i.e. $\dfrac{1}{4}$.

The equation of the normal is given by $y - y_1 = m(x - x_1)$

i.e. $y - 4 = \tfrac{1}{4}(x - 1) \quad \Rightarrow \quad 4y = x + 15$

2 Find the equation of the tangent to the curve $y = x^2 - 6x + 5$ at each of the points where the curve crosses the x-axis. Find also the coordinates of the point where these tangents meet.

The curve crosses the x-axis where $y = 0$

i.e. where $x^2 - 6x + 5 = 0 \Rightarrow (x - 5)(x - 1) = 0$

$\Rightarrow \quad x = 5$ and $x = 1$

Therefore the curve crosses the x-axis at $(5, 0)$ and $(1, 0)$.

$y = x^2 - 6x + 5 \quad \Rightarrow \quad \dfrac{dy}{dx} = 2x - 6$

At $(5, 0)$, the gradient of the tangent is given by $\dfrac{dy}{dx} = 10 - 6 = 4$

therefore the equation of this tangent is

$y - 0 = 4(x - 5) \quad \Rightarrow \quad y = 4x - 20$

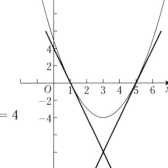

At $(1, 0)$ the gradient of the tangent is given by $\dfrac{dy}{dx} = 2 - 6 = -4$

Therefore the equation of the tangent is $y - 0 = -4(x - 1) \Rightarrow y + 4x = 4$

The two tangents meet at P, so at P,

$y + 4x = 4$ and $y - 4x = -20$

Solving these equations simultaneously gives $2y = -16 \Rightarrow y = -8$

Using $y = -8$ in $y + 4x = 4$ gives $-8 + 4x = 4 \Rightarrow x = 3$

Therefore the tangents meet at $(3, -8)$.

Exercise 8a

In each question from **1** to **6** find, at the given point,

(a) the equation of the tangent

(b) the equation of the normal.

1 $y = x^2 - 4$ where $x = 1$

2 $y = x^2 + 4x - 2$ where $x = 0$

3 $y = \dfrac{1}{x}$ where $x = -1$

4 $y = x^2 + 5$ where $x = 0$

5 $y = x^2 - 5x + 7$ where $x = 2$

6 $y = (x - 2)(x^2 - 1)$ where $x = -2$

7 Find the equation of the normal to the curve $y = x^2 + 4x - 3$ at the point where the curve cuts the y-axis.

8 Find the equation of the tangent to the curve $y = x^2 - 3x - 4$ at the point where this curve cuts the line $x = 5$

9 Find the equation of the tangent to the curve $y = (2x - 3)(x - 1)$ at each of the points where this curve cuts the x-axis. Find the point of intersection of these tangents.

10 Find the equation of the normal to the curve $y = x^2 - 6x + 5$ at each of the points where the curve cuts the x-axis.

11 Find the equation of the tangent to the curve $y = 3x^2 + 5x - 1$ at each of the points of intersection of the curve and the line $y = x - 1$

12 Find the equation of the tangent to the curve $y = x^2 + 5x - 3$ at the points where the line $y = x + 2$ crosses the curve.

13 Find the coordinates of the point on the curve $y = 2x^2$ where the gradient is 8.

Hence find the equation of the tangent to $y = 2x^2$ whose gradient is 8.

14 Find the coordinates of the point on the curve $y = 3x^2 - 1$ where the gradient of the tangent is 3.

15 Find the equation of the tangent to the curve $y = 4x^2 + 3x$ whose gradient is -1.

16 Find the equation of the normal to the curve $y = 2x^2 - 2x + 1$ whose gradient is $\frac{1}{2}$.

17 Find the value of k for which $y = 2x + k$ is a tangent to the curve $y = 2x^2 - 3$

18 Find the equation of the tangent to the curve $y = (x - 5)(2x + 1)$ that is parallel to the x-axis.

19 Find the coordinates of the point(s) on the curve $y = x^2 - 5x + 3$ where the gradient of the normal is $\frac{1}{3}$.

20 A curve has the equation $y = x^3 - px + q$. The tangent to this curve at the point $(2 -8)$ is parallel to the x-axis. Find the values of p and q.

Find also the coordinates of the other point where the tangent is parallel to the x-axis.

STATIONARY VALUES

For a function f, the derived function, $f'(x)$, expresses the rate at which $f(x)$ increases with respect to x.

At a particular point,

if $f'(x)$ is positive then $f(x)$ is increasing as x increases, whereas if $f'(x)$ is negative then $f(x)$ is decreasing as x increases.

There may be points where $f'(x)$ is zero, i.e. $f(x)$ is momentarily neither increasing nor decreasing with respect to x.

The value of $f(x)$ at such a point is called a *stationary value*

i.e. $f'(x) = 0$ $\Rightarrow$ $f(x)$ has a stationary value.

Look at the graph of the curve with equation $y = f(x)$

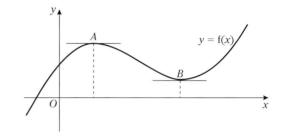

At A and B, $f(x)$, and therefore y, is neither increasing nor decreasing with respect to x. So the values of y at A and B are stationary values,

i.e. $\dfrac{dy}{dx} = 0$ $\Rightarrow$ y has a stationary value.

The point on a curve where y has a stationary value is called a *stationary point* and at any stationary point, the gradient of the tangent to the curve is zero, i.e. the tangent is parallel to the x-axis.

To sum up:

at a stationary point $\begin{cases} \textbf{\textit{y}, or f(\textit{x}) has a stationary value} \\[4pt] \dfrac{\textbf{d}\textit{\textbf{y}}}{\textbf{d}\textit{\textbf{x}}}\textbf{, or f}'(\textit{\textbf{x}})\textbf{, is zero} \\[4pt] \textbf{the tangent is parallel to the \textit{x}-axis.} \end{cases}$

Example 8b

Find the stationary values of the function $x^3 - 4x^2 + 7$.

If $f(x) = x^3 - 4x^2 + 7$

then $f'(x) = 3x^2 - 8x$

At stationary points, $f'(x) = 0$ i.e. $3x^2 - 8x = 0$

$\Rightarrow$ $x(3x - 8) = 0 \Rightarrow x = 0$ and $x = \frac{8}{3}$

Therefore there are stationary points where $x = 0$ and $x = \frac{8}{3}$

When $x = 0$, $f(x) = 0 - 0 + 7 = 7$

When $x = \frac{8}{3}$, $f(x) = \left(\frac{8}{3}\right)^3 - 4\left(\frac{8}{3}\right)^2 + 7 = -2\frac{13}{27}$

Therefore the stationary values of $x^3 - 4x^2 - 5$ are 7 and $-2\frac{13}{27}$.

Exercise 8b

Find the value(s) of x at which the following functions have stationary values.

1 $x^2 + 7$

2 $2x^2 - 3x - 2$

3 $x^3 - 4x^2 + 6$

4 $4x^3 - 3x - 9$

5 $x^3 - 2x^2 + 11$

6 $x^3 - 3x - 5$

Find the value(s) of x for which y has a stationary value.

7 $y = x^2 - 8x + 1$

8 $y = x + \dfrac{9}{x}$

9 $y = 2x^3 + x^2 - 8x + 1$

10 $y = 9x^3 - 25x$

11 $y = 2x^3 + 9x^2 - 24x + 7$

12 $y = 3x^3 - 12x + 1$

Find the coordinates of the stationary points on the following curves.

13 $y = \dfrac{x^2 + 9}{2x}$

14 $y = x^3 - 2x^2 + x - 7$

15 $y = (x - 3)(x + 2)$

16 $y = x^{\frac{3}{2}} - x^{\frac{1}{2}}$

17 $y = \sqrt{x} + \dfrac{1}{\sqrt{x}}$

18 $y = 8 + \dfrac{x}{4} + \dfrac{4}{x}$

TURNING POINTS

In the immediate neighbourhood of a stationary point a curve can have any one of the shapes shown in the diagram.

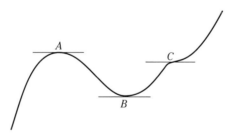

Moving through A from left to right, the curve is rising, then turns at A and begins to fall, i.e. the gradient changes from positive, to zero at A, and then becomes negative.

At A there is a *turning point*.

The value of y at A is called a *maximum value* and A is called a *maximum point*.

Moving through B from left to right the curve is falling, then turns at B and begins to rise, i.e. the gradient changes from negative, to zero at B, and then becomes positive.

At B there is a *turning point*.

The value of y at B is called a *minimum value* and B is called a *minimum point*.

The tangent is always horizontal at a turning point.

A maximum value of y is *not necessarily the greatest value of y overall*. The terms maximum and minimum apply only to the behaviour of the curve near a stationary point.

At C the curve does not turn. The gradient goes from positive, to zero at C and then becomes positive again, i.e. the gradient does not change sign at C. C is not a turning point, it is called a point of inflexion.

Distinguishing between turning points

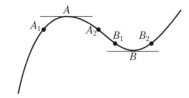

Method 1

This method compares the value of y at the stationary point with values of y at points on either side of, and near to, the stationary point.

For a maximum value, e.g. at A

y at $A_1 < y$ at A

y at $A_2 < y$ at A

For a minimum value, e.g. at B

y at $B_1 > y$ at B

y at $B_2 > y$ at B

Collecting these conclusions we have:

	Maximum	Minimum
y values on each side of the stationary point	both smaller	both larger

If y does not satisfy one of these conditions, there is not a turning point.

The points chosen on either side of the stationary point must be such that no *other* stationary point, nor any break in the graph, lies between them.

Method 2

This method looks at the sign of the gradient at points close to, and on either side of, the stationary point.

For a maximum point, A

$\dfrac{dy}{dx}$ at A_1 is +ve, $\dfrac{dy}{dx}$ at A_2 is −ve

For a minimum point, B

$\dfrac{dy}{dx}$ at B_1 is −ve, $\dfrac{dy}{dx}$ at B_2 is +ve

Collecting these conclusions in a table gives

Sign of $\dfrac{dy}{dx}$	Passing through maximum + 0 −	Passing through minimum − 0 +−
Gradient of tangent	/ — \	\ __ /

Method 3

This method uses the rate of change of $\dfrac{dy}{dx}$.

The rate of change of $\dfrac{dy}{dx}$ is $\dfrac{d}{dx}\left(\dfrac{dy}{dx}\right)$ which is written as $\dfrac{d^2y}{dx^2}$.

$\dfrac{d^2y}{dx^2}$ is the second derivative of y with respect to x.

For the maximum point A: at A_1 $\dfrac{dy}{dx}$ is +ve and at A_2 $\dfrac{dy}{dx}$ is −ve

so, passing through A, $\dfrac{dy}{dx}$ goes from + to −, i.e. $\dfrac{dy}{dx}$ decreases,

therefore the rate of change of $\dfrac{dy}{dx}$ is negative

$\Rightarrow$ at A, $\dfrac{d^2y}{dx^2}$ is negative.

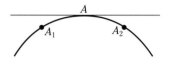

For the minimum point B: at B_1 $\dfrac{dy}{dx}$ is −ve and at B_2 $\dfrac{dy}{dx}$ is +ve

so, passing through B, $\dfrac{dy}{dx}$ goes from − to +, i.e. $\dfrac{dy}{dx}$ increases,

therefore the rate of change of $\dfrac{dy}{dx}$ is positive

$\Rightarrow$ at B, $\dfrac{d^2y}{dx^2}$ is positive.

Summing up method 3 we have:

	Maximum	Minimum
Sign of $\dfrac{d^2y}{dx^2}$	negative (or zero)	positive (or zero)

There are stationary points where $\dfrac{d^2y}{dx^2} = 0$, but you will not be examined on these.

Examples 8c

1 Find the coordinates of the stationary points on the curve $y = 4x^3 + 3x^2 - 6x - 1$ and determine the nature of each one.

$y = 4x^3 + 3x^2 - 6x - 1$

$\Rightarrow \quad \dfrac{dy}{dx} = 12x^2 + 6x - 6$

At stationary points, $\dfrac{dy}{dx} = 0$

i.e. $12x^2 + 6x - 6 = 0$

$\Rightarrow \quad 6(2x - 1)(x + 1) = 0$

$\therefore$ there are stationary points where $x = \tfrac{1}{2}$ and $x = -1$

When $x = \tfrac{1}{2}$, $y = -2\tfrac{3}{4}$ and when $x = -1$, $y = 4$

i.e. the stationary points are $(\tfrac{1}{2}, -2\tfrac{3}{4})$ and $(-1, 4)$.

Differentiating $\dfrac{dy}{dx}$ with respect to x gives

$\dfrac{d^2y}{dx^2} = 24x + 6$

When $x = \frac{1}{2}$

$\dfrac{d^2y}{dx^2} = 12 + 6$ which is positive

$\Rightarrow \quad (\frac{1}{2}, -2\frac{3}{4})$ is a minimum point.

When $x = -1$

$\dfrac{d^2y}{dx^2} = -24 + 6$ which is negative

$\Rightarrow \quad (-1, 4)$ is a maximum point.

2 Show that $3x^4 - 8x^3 + 6x^2 - 3$ has a stationary value when $x = 0$ and determine its nature.

$f(x) = 3x^4 - 8x^3 + 6x^2 - 3 \quad \Rightarrow \quad f'(x) = 12x^3 - 24x^2 + 12x$

At stationary values $f'(x) = 0$

i.e. $12x^3 - 24x^2 + 12x = 0 \quad \Rightarrow \quad 12x(x^2 - 2x + 1) = 0$

$\Rightarrow \quad 12x(x - 1)(x - 1) = 0$

$\Rightarrow \quad x = 0 \text{ or } x = 1$

therefore there is a stationary value when $x = 0$

Differentiating $f'(x)$ with respect to x gives

$f''(x) = 36x^2 - 48x + 12 = 12(3x^2 - 4x + 1)$ $f''(x)$ is the notation for $\dfrac{d(f'(x))}{dx}$

When $x = 0$, $f''(x) = 12$ which is positive

$\Rightarrow \quad f(x) = -3$ is a minimum value.

3 The function f is given by $x \mapsto ax^2 + bx + c, x \in \mathbb{R}$
$f'(x) = 4x + 2$ and f has a stationary value of 1. Find the values of a, b and c.

$f(x) = ax^2 + bx + c \quad \Rightarrow \quad f'(x) = 2ax + b$

But we know that $f'(x) = 4x + 2$

$\therefore \quad 2ax + b$ is identical to $4x + 2$

i.e. $a = 2$ and $b = 2$

The stationary value of $f(x)$ occurs when $f'(x) = 0$

i.e. when $4x + 2 = 0 \quad \Rightarrow \quad x = -\frac{1}{2}$

the stationary value of $f(x)$ is $\quad 2\left(-\frac{1}{2}\right)^2 + 2\left(-\frac{1}{2}\right) + c = -\frac{1}{2} + c$

But the stationary value of $f(x)$ is also 1,

$\therefore \quad -\frac{1}{2} + c = 1 \quad \Rightarrow \quad c = \frac{3}{2}$

4 A cylinder has a radius r metres and a height h metres. The sum of the radius and height is 2 m. The volume of the cylinder is $V\,\mathrm{cm}^3$.

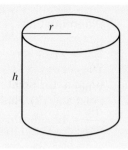

(a) Show that $V = \pi r^2(2 - r)$

(b) Find the maximum value of V.

(a) $V = \pi r^2 h$ and $r + h = 2$

$\therefore \qquad V = \pi r^2(2 - r) = \pi(2r^2 - r^3)$

(b) V is maximum when $\dfrac{\mathrm{d}V}{\mathrm{d}r} = 0$

i.e. $\pi(4r - 3r^2) = 0 \ \Rightarrow \ \pi r(4 - 3r) = 0$

Therefore there are stationary values of V when $r = 0$ and $r = \frac{4}{3}$

It is obvious that, when $r = 0$, $V = 0$ and no cylinder exists, so we check the

sign of $\dfrac{\mathrm{d}^2V}{\mathrm{d}r^2}$ only for $r = \frac{4}{3}$

$\dfrac{\mathrm{d}^2V}{\mathrm{d}r^2} = \pi(4 - 6r)$ which is negative when $r = \frac{4}{3}$

Therefore the maximum value of V occurs when $r = \frac{4}{3}$

The maximum value of V is $\pi \left(\frac{4}{3}\right)^2 \left(2 - \frac{4}{3}\right) = \frac{32}{27}\,\pi$

Exercise 8c

Find the coordinates of the maximum and minimum points on the following curves and determine their nature.

1 $y = 2x - x^2$

2 $y = 3x - x^3$

3 $y = \dfrac{9}{x} + x$

4 $y = x^2(x - 5)$

5 $y = x^2$

6 $y = x + \frac{1}{2}x^2$

7 $y = 2x^2 - x^4$

8 $y = x^4 - 32x$

9 $y = (2x + 1)(x - 3)$

10 $y = x^5 - 5x$

11 $y = x^2(x^2 - 8)$

12 $y = x^2 + \dfrac{16}{x^2}$

Find the stationary value(s) of each of the following functions.

13 $x + \dfrac{1}{x}$

14 $3 - x + x^2$

15 $4x^3 - x^4$

16 $8 - x^3$

17 $x^3 + 7$

18 $x^2(3x^2 - 2x - 3)$

19 Show that the curve with equation
$y = x^5 + x^3 + 4x - 3$ has no stationary
points.

20 The curve $y = ax^2 + bx + c$ crosses the
y-axis at the point $(0, 3)$ and has a stationary
point at $(1, 2)$. Find the values of a, b and c.

21 The gradient of the tangent to the curve
$y = px^2 - qx - r$ at the point $(1, -2)$ is 1.
If the curve crosses the x-axis where $x = 2$,
find the values of p, q and r. Find the other
point of intersection with the x-axis and
sketch the curve.

22 $y = ax^2 + bx + c$. The line $y = 2x$ is a tangent
to the curve at the point $(0, 6)$. The turning
point on the curve occurs where $x = -2$.
Find the values of a, b and c.

23

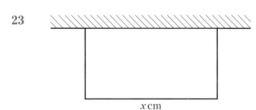

x cm

A wire, 80 cm long, is bent to form three
sides of a rectangle against a fixed wall as
shown in the diagram. The area enclosed is
A cm^2.

(a) Show that $A = \frac{1}{2}x(80 - x)$

(b) Find the value of x for which A has its
maximum value.

24

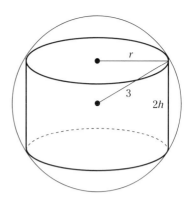

The diagram shows a cylinder of radius r cm
and height $2h$ cm cut from a solid sphere of
radius 3 cm. The volume of the cylinder is
V cm^3.

(a) Show that $r = \sqrt{9 - h^2}$ and hence that
$V = 2\pi h(9 - h^2)$

(b) Given that h varies, find the stationary
value of V and determine its nature.

25

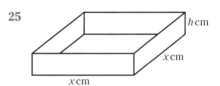

h cm
x cm
x cm

An open box in the shape of a cuboid has a
square base of side x cm and height h cm. The
volume, V cm^3, inside the box is 4000 cm^3.

(a) Show that $h = \dfrac{4000}{x^2}$

(b) Given that x varies, show that
$\dfrac{\mathrm{d}h}{\mathrm{d}x} = -1000$ when $x = 2$

Mixed exercise 8

1 Find the gradient of the curve with equation
$y = 6x^2 - x$ at the point where $x = 1$
Find the equation of the tangent at this point.
Where does this tangent meet the line $y = 2x$?

2 Find the equation of the normal to the curve
$y = 1 - x^2$ at the point where the curve
crosses the positive x-axis. Find also the
coordinates of the point where the normal
meets the curve again.

3 Find the coordinates of the points on the
curve $y = x^3 + 3x$ where the gradient is 15.

4 Find the equations of the tangents to the
curve $y = x^3 - 6x^2 + 12x + 2$ that are parallel
to the line $y = 3x$

5 Find the equation of the normal to the curve
$y = x^2 - 6$ that is parallel to the line
$x + 2y - 1 = 0$

6 Locate the turning points on the curve
$y = x(x^2 - 12)$, determine their nature and
draw a rough sketch of the curve.

7 Find the stationary values of the function
$x + \dfrac{1}{x}$ and sketch the function.

8

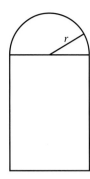

A rectangle of width $2r$ m has a semicircle of radius r m joined to one side as shown in the diagram.

The perimeter of the shape is 7 m.

(a) Show that the height, h m, of the rectangle is given by $2h = 7 - r(2 + \pi)$

(b) Show that the area of the shape is a maximum when $r = \dfrac{7}{4 + \pi}$

Summary 2

FUNCTIONS

A function f is a rule that maps a number x to another single number $f(x)$. The domain of a function is the set of input numbers, i.e. the set of values of x. When $x \in \mathbb{R}$, x can have any value.

The range of a function is the set of output values, i.e. the set of values of $f(x)$.

The general form of a quadratic function is

$$f(x) = ax^2 + bx + c \text{ where } a \neq 0$$

If $a > 0$, $f(x)$ has a minimum value where

$$x = \frac{-b}{2a}$$

If $a < 0$, $f(x)$ has a maximum value where

$$x = \frac{-b}{2a}$$

The general form of a polynomial function is

$$f(x) = a_n x^n + a_{x-1} x^{n-1} + \ldots + a_o$$

where n is a positive integer and a_n, a_{n-1}, are rational constants.

The general form of a rational function is $\dfrac{f(x)}{g(x)}$

where $f(x)$ and $g(x)$ are polynomials.

The function that maps the output of f to its input is called the inverse function of f, and is denoted by f^{-1}, i.e. $f^{-1} : f(x) \mapsto x$. The range of f is the domain of f^{-1}.

When the mapping is not one–one, the function does not have an inverse.

When a function g is applied to a function f, the result is a composite function denoted by gf.

INEQUALITIES

If $a > b$ then $a + k > b + k$ for all values of k

$\qquad ak > bk$ for all positive values of k

CURVES

A chord is a straight line joining two points on a curve.

A tangent to a curve is a line that touches the curve at one point, called the point of contact.

A normal to a curve is the line perpendicular to a tangent and through its point of contact.

The gradient of a curve at a point on the curve is the gradient of the tangent at that point.

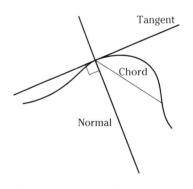

INTERSECTION OF A LINE AND A CURVE

The coordinates of the points of intersection of a line and a curve can be found by solving the equations of the line and the curve simultaneously.

When there are no real roots the line does not intersect the curve.

When a root is repeated, the line is a tangent to the curve.

DIFFERENTIATION

Differentiation is the process of finding a general expression for the gradient of a curve at any point on the curve.

This general expression is called the gradient function, or the derived function or the derivative.

The derivative is denoted by $\dfrac{dy}{dx}$ or by $f'(x)$

When $y = x^n$, $\dfrac{dy}{dx} = nx^{n-1}$

When $y = ax^n$, $\dfrac{dy}{dx} = anx^{n-1}$

When $y = c$, $\dfrac{dy}{dx} = 0$

THE CHAIN RULE

When $y = fg(x)$, where $u = g(x)$ so $y = f(u)$

then $\dfrac{dy}{dx} = \dfrac{dy}{du} \times \dfrac{du}{dx}$

STATIONARY VALUES

A stationary value of $f(x)$ is its value where $f'(x) = 0$

The point on the curve $y = f(x)$ where $f(x)$ has a stationary value is called a stationary point.
At all stationary points, $\dfrac{dy}{dx} = 0$.

TURNING POINTS

A, B and C are stationary points on the curve $y = f(x)$.
The points A and B are called turning points.
The points C is called a point of inflexion.

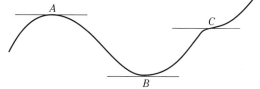

At A, $f(x)$ has a maximum value and A is called a maximum point.

At B, $f(x)$ has a minimum value and B is called a minimum point.

There are three methods for distinguishing stationary points.

	Max	Min
1 Find value of y on each side of stationary value	Both smaller	Both larger
2 Find sign of $\dfrac{dy}{dx}$ on each side of stationary value Gradient	$+\quad 0 \quad -$ $/\ \ \overline{\ }\ \ \backslash$	$-\quad 0 \quad +$ $\backslash\ \_\ /$
3 Find sign of $\dfrac{d^2y}{dx^2}$ at stationary value	$-$ve (or 0)	$+$ve (or 0)

Method 3 is the easiest to apply but it fails if $\dfrac{d^2y}{dx^2}$ is zero. However, you will not be examined on this.

Summary exercise 2

1 The function f is defined by

$$f : x \mapsto 3x - 2 \text{ for } x \in \mathbb{R}$$

(i) Sketch, in a single diagram, the graphs of $y = f(x)$ and $y = f^{-1}(x)$, making clear the relationship between the two graphs. [2]

The function g is defined by

$$g : x \mapsto 6x - x^2 \text{ for } x \in \mathbb{R}$$

(ii) Express $gf(x)$ in terms of x, and hence show that the maximum value of $gf(x)$ is 9. [5]

The function h is defined by

$$h : x \mapsto 6x - x^2 \text{ for } x \geq 3$$

(iii) Express $6x - x^2$ in the form $a - (x - b)^2$, where a and b are positive constants. [2]

(iv) Express $h^{-1}(x)$ in terms of x. [3]

Cambridge, Paper 1 Q10 N08

2 The equation of a curve is $y = \dfrac{12}{x^2 + 3}$

(i) Obtain an expression for $\dfrac{dy}{dx}$ [2]

(ii) Find the equation of the normal to the curve at the point $P(1, 3)$. [3]

(iii) A point is moving along the curve in such a way that the x-coordinate is increasing at a constant rate of 0.012 units per second. Find the rate of change of the y-coordinate as the point passes through P. [2]

Cambridge, Paper 11 Q7 N09

3 The function f is defined by

$$f : x \mapsto 2x^2 - 8x + 11 \text{ for } x \in \mathbb{R}$$

(i) Express $f(x)$ in the form $a(x + b)^2 + c$, where a, b and c are constants. [3]

(ii) State the range of f. [1]

(iii) Explain why f does not have an inverse. [1]

The function g is defined by

$$g : x \mapsto 2x^2 - 8x + 11 \text{ for } x \leq A, \text{ where } A \text{ is a constant.}$$

(iv) State the largest value of A for which g has an inverse. [1]

(v) When A has this value, obtain an expression, in terms of x, for $g^{-1}(x)$ and state the range of g^{-1}. [4]

Cambridge, Paper 1 Q11 N07

4

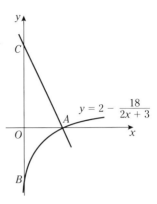

The diagram shows part of the curve $y = 2 - \dfrac{18}{2x + 3}$, which crosses the x-axis at A and the y-axis at B. The normal to the curve at A crosses the y-axis at C.

(i) Show that the equation of the line AC is $9x + 4y = 27$ [6]

(ii) Find the length of BC. [2]

Cambridge, Paper 11 Q7 J10

5 The function f is defined by

$$f : x \mapsto 2x^2 - 12x + 7 \text{ for } x \in \mathbb{R}$$

(i) Express $f(x)$ in the form $a(x + b)^2 - c$ [3]

(ii) State the range of f. [1]

(iii) Find the set of values of x for which $f(x) < 21$ [3]

The function g is defined by

$$g : x \mapsto 2x + k \text{ for } x \in \mathbb{R}$$

(iv) Find the value of the constant k for which the equation $gf(x) = 0$ has two equal roots. [4]

Cambridge, Paper 11 Q9 J10

6 The function f is defined by

$$f : x \mapsto 2x^2 - 12x + 13 \text{ for } 0 \leq x \leq A, \text{ where } A \text{ is a constant.}$$

(i) Express $f(x)$ in the form $a(x + b)^2 + c$, where a, b and c are constants. [3]

(ii) State the value of A for which the graph of $y = f(x)$ has a line of symmetry. [1]

(iii) When A has this value, find the range of f. [2]

The function g is defined by

$$g : x \mapsto 2x^2 - 12x + 13 \text{ for } x \geq 4$$

(iv) Explain why g has an inverse. [1]

(v) Obtain an expression, in terms of x, for $g^{-1}(x)$. [3]

Cambridge, Paper 1 Q10 J09

7 The equation of a curve C is $y = 2x^2 - 8x + 9$ and the equation of a line L is $x + y = 3$

  (i) Find the x-coordinates of the points of intersection of L and C.   [4]

  (ii) Show that one of these points is also the stationary point of C.   [3]

    Cambridge, Paper 1 Q4 J08

8 The function f is such that $f(x) = (3x + 2)^3 - 5$ for $x \geqslant 0$

  (i) Obtain an expression for $f'(x)$ and hence explain why f is an increasing function. [3]

  (ii) Obtain an expression for $f^{-1}(x)$ and state the domain of f^{-1} [4]

    Cambridge, Paper 1 Q6 J08

9

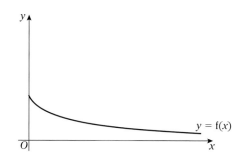

The diagram shows the graph of $y = f(x)$, where $f : x \mapsto \dfrac{6}{2x + 3}$ for $x \geqslant 0$

  (i) Find an exprcssion, in terms of x, for $f'(x)$ and explain how your answer shows that f is a decreasing function.   [3]

  (ii) Find an expression, in terms of x, for $f^{-1}(x)$ and find the domain of f^{-1} [4]

  (iii) Copy the diagram and, on your copy, sketch the graph of $y = f^{-1}(x)$, making clear the relationship between the graphs.   [2]

The function g is defined by $g : x \mapsto \frac{1}{2}x$ for $x \geqslant 0$

  (iv) Solve the equation $fg(x) = \frac{3}{2}$ [3]

    Cambridge, Paper 1 Q11 J07

10

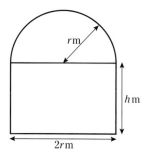

The diagram shows a glass window consisting of a rectangle of height h m and width $2r$ m and a semicircle of radius r m. The perimeter of the window is 8 m.

  (i) Express h in terms of r.   [2]

  (ii) Show that the area of the window, A m², is given by

$$A = 8r - 2r^2 - \tfrac{1}{2}\pi r^2$$ [2]

Given that r can vary,

  (iii) find the value of r for which A has a stationary value   [4]

  (iv) determine whether this stationary value is a maximum or a minimum.   [3]

    Cambridge, Paper 1 Q8 J04

11 The functions f and g are defined as follows:

$f : x \mapsto x^2 - 2x, \, x \in \mathbb{R}$

$g : x \mapsto 2x + 3, \, x \in \mathbb{R}$

  (i) Find the set of values of x for which $f(x) > 15$ [3]

  (ii) Find the range of f and state, with a reason, whether f has an inverse.   [4]

  (iii) Show that the equation $gf(x) = 0$ has no real solutions.   [3]

  (iv) Sketch, in a single diagram, the graphs of $y = g(x)$ and $y = g^{-1}(x)$, making clear the relationship between the graphs. [2]

    Cambridge, Paper 1 Q10 J04

12 Determine the set of values of the constant k for which the line $y = 4x + k$ does not intersect the curve $y = x^2$ [3]

    Cambridge, Paper 1 Q1 N07

13 Find the set of values of k for which the line $y = kx - 4$ intersects the curve $y = x^2 - 2x$ at two distinct points.   [4]

    Cambridge, Paper 1 Q2 J09

14 Find the value of the constant c for which the line $y = 2x + c$ is a tangent to the curve $y^2 = 4x$ [4]

    Cambridge, Paper 1 Q1 J07

9 Trigonometry 1

After studying this chapter you should be able to

- sketch and use graphs of the sine, cosine and tangent functions for angles of any size and using either degrees or radians
- use the exact values of the sine, cosine and tangent of 30°, 45°, 60° and related angles.

THE TRIGONOMETRIC FUNCTIONS

The general definition of an angle

A line can rotate from its initial position OP_0 about the point O to any other position OP.

The amount of rotation is measured by the angle between OP_0 and OP, i.e.

   an angle is a measure of the rotation of a line about a fixed point.

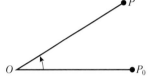

The anticlockwise sense of rotation is taken as positive and clockwise rotation is negative.

It follows that an angle formed by the anticlockwise rotation of OP is a positive angle.

The rotation of OP can be more than one revolution, so an angle can be as big as we want to make it.

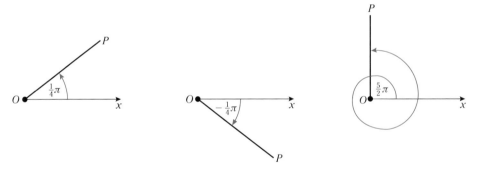

If θ is any angle, then θ can be measured either in degrees or in radians and in either case θ can take all real values.

The trigonometric functions

Since angles are no longer restricted in size, we also need general definitions for the sine, cosine and tangent of an angle that are valid for angles of all values.

If OP is drawn on x- and y-axes as shown
and, for all values of θ, the length of OP is r and the coordinates of P are (x, y),
then the sine, cosine and tangent functions are defined as follows.

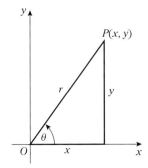

$$\sin\theta = \frac{y}{r}$$

$$\cos\theta = \frac{x}{r}$$

$$\tan\theta = \frac{y}{x}$$

THE TRIGONOMETRIC RATIOS OF 30°, 45°, 60°

The sine, cosine and tangent of 30°, 45° and 60° can be expressed exactly in surd form and are worth remembering.

This triangle shows that

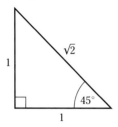

$$\sin 45° = \frac{1}{\sqrt{2}}$$

$$\cos 45° = \frac{1}{\sqrt{2}}$$

$$\tan 45° = 1$$

And this triangle gives

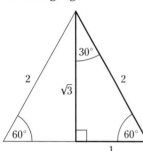

$$\sin 60° = \frac{\sqrt{3}}{2}, \sin 30° = \frac{1}{2}$$

$$\cos 60° = \frac{1}{2}, \cos 30° = \frac{\sqrt{3}}{2}$$

$$\tan 60° = \sqrt{3}, \tan 30° = \frac{1}{\sqrt{3}}$$

THE SINE FUNCTION

From the definition $f(\theta) = \sin \theta$, and measuring θ in radians, we can see that:

for $0 \leqslant \theta \leqslant \frac{1}{2}\pi$, OP is in the first quadrant;
y is positive and increases in value from
0 to r as θ increases from 0 to $\frac{1}{2}\pi$.

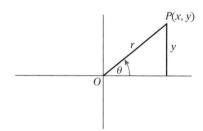

Now r is always positive, so $\sin \theta$ increases from 0 to 1.

For $\frac{1}{2}\pi \leqslant \theta \leqslant \pi$, OP is the second quadrant;
again y is positive but decreases in value from r to 0,

so $\sin \theta$ decreases from 1 to 0.

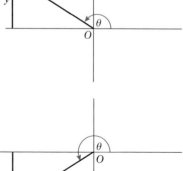

For $\pi \leqslant \theta \leqslant \frac{3}{2}\pi$, OP is the third quadrant;
y is negative and decreases from 0 to $-r$,

so $\sin \theta$ decreases from 0 to -1.

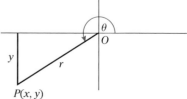

For $\frac{3}{2}\pi \leqslant \theta \leqslant 2\pi$, OP is the fourth quadrant; y is still negative but increases from $-r$ to 0,

so $\sin\theta$ increases from -1 to 0.

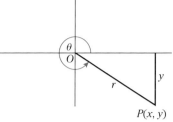

For $\theta > 2\pi$, the cycle repeats itself as OP travels round the quadrants again.
For negative values of θ, OP rotates clockwise round the quadrants in the order 4th, 3rd, 2nd, 1st, etc. So $\sin\theta$ decreases from 0 to -1, then increases to 0 and on to 1 before decreasing to zero and repeating the pattern.

This shows that $\sin\theta$ is positive for $0 < \theta < \pi$ and negative when $\pi < \theta < 2\pi$

Also, $\sin\theta$ varies in value between -1 and 1 and the pattern repeats itself every revolution.

A plot of the graph of $f(\theta) = \sin\theta$ confirms these observations.

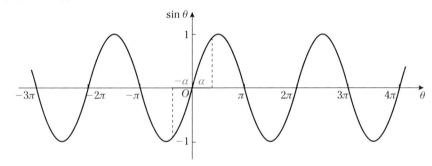

A graph of this shape is called a *sine wave* and shows the following properties of the sine function.

The curve is continuous (i.e. it has no breaks).

$$-1 \leqslant \sin\theta \leqslant 1$$

The shape of the curve from $\theta = 0$ to $\theta = 2\pi$ is repeated for each complete revolution. Any function with a repetitive pattern is called *periodic* or *cyclic*. The width of the repeating pattern, as measured on the horizontal scale, is called the *period*.

The period of the sine function is 2π.

Other properties of the sine function shown by the graph are as follows.

$\sin\theta = 0$ when $\theta = n\pi$ where n is an integer.

The curve has rotational symmetry about the origin so, for any angle α

$$\sin(-\alpha) = -\sin\alpha, \quad \text{e.g.} \quad \sin(-30°) = -\sin 30° = -\frac{1}{2}$$

An enlarged section of the graph for $0 \leqslant \theta \leqslant 2\pi$ shows further relationships.

The curve is symmetrical about the line $\theta = \frac{1}{2}\pi$, so

$$\sin(\pi - \alpha) = \sin\alpha$$

e.g. $\sin 130° = \sin(180° - 130°) = \sin 50°$

The curve has rotational symmetry about $\theta = \pi$

so $\sin(\pi + \alpha) = -\sin\alpha$

and $\sin(2\pi - \alpha) = -\sin\alpha$

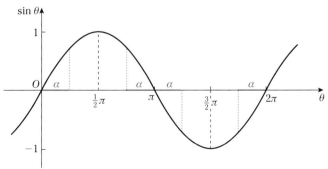

The graph of f(θ) = sin $k\theta$

The following table gives pairs of corresponding values of θ and f(θ) = sin 2θ

θ	0	$\dfrac{\pi}{4}$	$\dfrac{\pi}{2}$	$\dfrac{3\pi}{4}$	π	$\dfrac{5\pi}{4}$	$\dfrac{3\pi}{2}$	$\dfrac{7\pi}{4}$	2π
2θ	0	$\dfrac{\pi}{2}$	π	$\dfrac{3\pi}{2}$	2π	$\dfrac{5\pi}{2}$	3π	$\dfrac{7\pi}{2}$	4π
f(θ)	0	1	0	-1	0	1	0	-1	0

Plotting these points gives this graph.

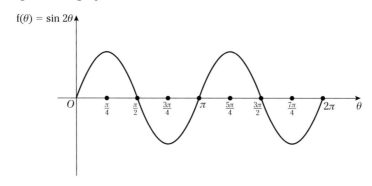

The graph has the following properties.

(a) f(θ) = sin 2θ is a cyclic and its period is $\pi \left(\text{i.e. } \dfrac{2\pi}{2}\right)$.

(b) The greatest and least values are 1 and -1.

(c) The shape is a sine wave.

(d) Within the range $0 \leqslant \theta \leqslant 2\pi$ there are two complete curve patterns compared with only one for the basic function f(θ) $\equiv$ sin θ, i.e. the complete *cycle* appears with twice the frequency.

When the graph of f(θ) $\equiv$ sin 3θ is plotted we find that the function is cyclic with a period $\dfrac{2\pi}{3}$, so that three complete cycles occur between 0 and 2π. In fact, the graph of the function f(θ) $\equiv$ sin $k\theta$ is a sine wave with a period of $\dfrac{2\pi}{k}$ and a frequency k times that of f(θ) $\equiv$ sin θ

The graph of f(θ) = sin (θ − α)

For the function f(θ) = sin ($\theta - \alpha$) where α is a positive constant, we have

f(θ) = 0   when sin ($\theta - \alpha$) = 0, π, 2π, 3π, ...

i.e.   when $\theta = 0 + \alpha$, $\pi + \alpha$, $2\pi + \alpha$, $3\pi + \alpha$, ...

f(θ) = 1   when sin ($\theta - \alpha$) = $\dfrac{\pi}{2}$, $\dfrac{5\pi}{2}$, ...

i.e.   when $\theta = \dfrac{\pi}{2} + \alpha$, $\dfrac{5\pi}{2} + \alpha$, ...

f(θ) = -1 when sin ($\theta - \alpha$) = $\dfrac{3\pi}{2}$, $\dfrac{7\pi}{2}$, ...

i.e.   when $\theta = \dfrac{3\pi}{2} + \alpha$, $\dfrac{7\pi}{2} + \alpha$, ...

So when $\alpha = \dfrac{\pi}{4}$, the graph of

$f(\theta) = \sin\left(\theta - \dfrac{\pi}{4}\right)$ is the same shape
as the graph of $y = \sin\theta$ but moved
by $\dfrac{\pi}{4}$ along in the *positive* direction of
the *x*-axis.

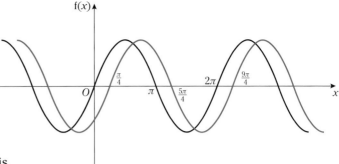

So for any positive value of α, the graph
of $f(\theta) = \sin(\theta - \alpha)$ is the same shape
as the graph of $y = \sin\theta$ but moved by
α along in the *positive* direction of the *x*-axis.

When $f(\theta) = \sin(\theta + \alpha)$, the graph is the same shape as graph of $y = \sin\theta$ but moved by α along in
the *negative* direction of the *x*-axis.

The graph of $f(\theta) = k + \sin\theta$

For any graph, $y = f(x)$, the graph of $y = f(x) + k$ where k is a constant, has the same shape as the
graph of $y = f(x)$ but moved by k units in the direction of the positive *y*-axis. When k is negative, the
movement is in the opposite direction.

The diagram shows the graphs of $f(\theta) = 0.5 + \sin\theta$ and $f(\theta) = -0.7 + \sin\theta$

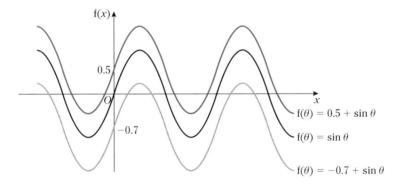

The graphs of $f(\theta) = -\sin\theta$ and $f(\theta) = \sin(-\theta)$

Comparing the equation $y = f(x)$ [1]

with the equation $y = -f(x)$ [2]

shows that, for the same value of *x*, the value of *y* in [2] is minus the
value of *y* in [1].

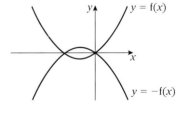

So the curves $y = f(x)$ and $y = -f(x)$ taken together are symmetrical
about the *x*-axis.

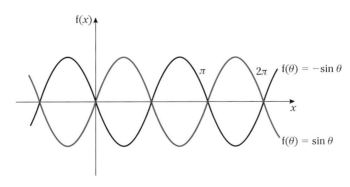

Comparing the equation $y = f(x)$ [1]

with the equation $y = f(-x)$ [2]

shows that, for the same value of y, the value of x in [2] is minus the value of x in [1].

So the curves $y = f(x)$ and $y = f(-x)$ taken together are symmetrical about the y-axis.

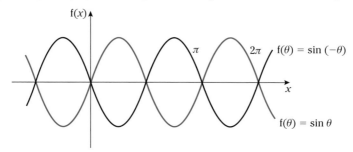

The Graph of $f(\theta) = k \sin \theta$

Comparing the equation $y = f(x)$ [1]

with the equation $y = k\,f(x)$ [2]

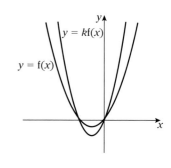

shows that, for the same value of x, the value of y in [2] is k times the value of y in [1].

So the curve $y = k\,f(x)$ is the curve $y = f(x)$ stretched by k units parallel to the y-axis.

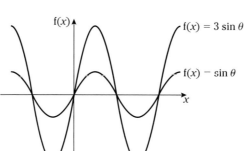

These facts about the graph $y = \sin \theta$ also apply to $y = \cos \theta$ and $y = \tan \theta$

Examples 9a

1 Find the exact value of $\sin \frac{4}{3}\pi$.

$$\sin \tfrac{4}{3}\pi = \sin\left(\pi + \tfrac{1}{3}\pi\right) = -\sin \tfrac{1}{3}\pi = -\frac{\sqrt{3}}{2}$$

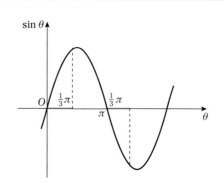

2   Sketch the graph of $y = 2 - \frac{1}{2} \sin \theta$ for $0 \leqslant \theta \leqslant 2\pi$.

Start with the graph of $y = \frac{1}{2} \sin \theta$, then the graph of $y = -\frac{1}{2} \sin \theta$. Finally move the graph 2 units up the y-axis.

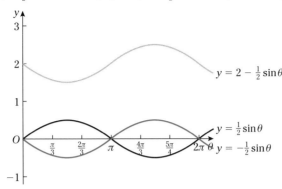

3   Sketch the graphs of $y = 1 + \sin \theta$ and $y = \theta$ for $-\pi \leqslant \theta \leqslant \pi$.

Hence state the number of solutions of the equation $\theta = 1 + \sin \theta$.

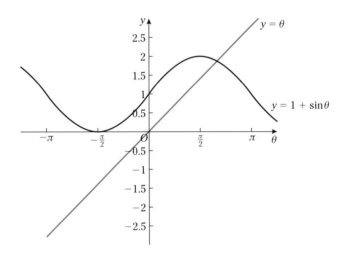

There is one solution.   There is only one point of intersection.

Exercise 9a

Find the exact value of

1   $\sin 120°$

2   $\sin -2\pi$

3   $\sin 300°$

4   $\sin -210°$

5   Write down all the values of θ between 0 and 6π for which $\sin \theta = 1$

6   Write down all the values of θ between 0 and -4π for which $\sin \theta = -1$

Express in terms of the sine of an acute angle

7   $\sin 125°$

8   $\sin 290°$

9   $\sin -120°$

10   $\sin \frac{7}{6}\pi$

Sketch each of the following curves for values of θ in the range $0 \leqslant \theta \leqslant 3\pi$

11   $y = \sin 2\theta$

12   $y = \sin (-\theta)$

13   $y = \sin (\pi - \theta)$

14   $y = -\sin \theta$

15   $y = 1 - \sin \theta$

16   $y = -2 + \sin \theta$

17   $3 - 4 \sin \theta$

18   $2 \sin 3\theta$

THE COSINE FUNCTION

For any position of P, $\cos \theta = \dfrac{x}{r}$

When P is in the first quadrant, x decreases from r to 0 as θ increases, so $\cos \theta$ decreases from 1 to 0.

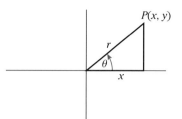

When P is in the second quadrant, x decreases from 0 to $-r$, so $\cos \theta$ decreases from 0 to -1.

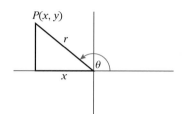

When P is in the third quadrant, $\cos \theta$ increases from -1 to 0,

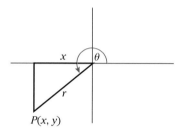

and when P is in the fourth quadrant, $\cos \theta$ increases from 0 to 1.

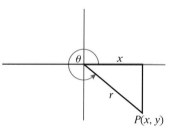

The cycle then repeats itself, and we get this graph of $\mathrm{f}(\theta) = \cos \theta$

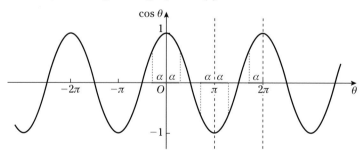

The properties of this graph are as follows.

The curve is continuous.

$$-1 \leqslant \cos \theta \leqslant 1$$

It is periodic with a period of 2π.

It is the same shape as the sine wave but is moved a distance $\frac{1}{2}\pi$ to the left. Such a move of a sine wave is called a *phase shift*.

$\cos \theta = 0$ when $\theta = \ldots -\frac{1}{2}\pi,\ \frac{1}{2}\pi,\ \frac{3}{2}\pi,\ \frac{5}{2}\pi, \ldots$

The curve is symmetrical about $\theta = 0$, so $\cos(-\alpha) = \cos \alpha$

The curve has rotational symmetry about $\theta = \frac{1}{2}\pi$, so

$$\cos(\pi - \alpha) = -\cos\alpha$$

Further considerations of symmetry show that

$$\cos(\pi + \alpha) = -\cos\alpha \quad \text{and} \quad \cos(2\pi - \alpha) = \cos\alpha$$

Exercise 9b

1 Write in terms of the cosine of an acute angle

   (a) $\cos 123°$ (b) $\cos 250°$

   (c) $\cos(-20°)$ (d) $\cos(-154°)$

2 Find the exact value of

   (a) $\cos 150°$ (b) $\cos\frac{3}{2}\pi$

   (c) $\cos\frac{5}{4}\pi$ (d) $\cos 6\pi$

3 *Sketch* each of the following curves.

   (a) $y = \cos(\theta + \pi)$ (b) $y = \cos\left(\theta - \frac{1}{3}\pi\right)$

   (c) $y = \cos(-\theta)$

4 Sketch the graph of $y = 3\cos 2\theta$ for $0 \leqslant \theta \leqslant \pi$.

5 On the same set of axes, sketch the graphs $y = \cos\theta$ and $y = 3\cos\theta$ for $0 \leqslant \theta \leqslant 2\pi$

6 On the same set of axes, sketch the graphs $y = \cos\theta$ and $y = \cos 3\theta$ for $0 \leqslant \theta \leqslant 2\pi$

7 Sketch the graphs of $f(\theta) = \cos 4\theta$ for $0 \leqslant \theta \leqslant \pi$. Hence find the values of θ in this range for which $f(\theta) = 0$

8 Sketch, for $0 \leqslant \theta \leqslant 2\pi$ the graphs of

   (a) $5 - 4\cos\theta$ (b) $2 + 3\cos\theta$

   (c) $\cos 4\theta$

9 Sketch the graph of $y = \cos\left(\theta - \frac{1}{4}\pi\right)$ for values of θ between $-\pi$ and π. Use the sketch to find the values of θ in this range for which

   (a) $\cos\left(\theta - \frac{1}{4}\pi\right) = 1$

   (b) $\cos\left(\theta - \frac{1}{4}\pi\right) = -1$

   (c) $\cos\left(\theta - \frac{1}{4}\pi\right) = 0$

THE TANGENT FUNCTION

For any position of P, $\tan\theta = \dfrac{y}{x}$

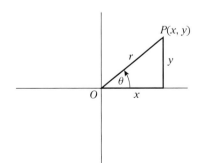

As OP rotates through the first quadrant, x decreases from r to 0, while y increases from 0 to r. This means that the fraction $\dfrac{y}{x}$ increases from 0 to very large values indeed. In fact, as $\theta \to \frac{1}{2}\pi$, $\tan\theta \to \infty$.

Looking at the behaviour of $\dfrac{y}{x}$ in the other quadrants shows that in the second quadrant, $\tan\theta$ is negative and increases from $-\infty$ to 0; in the third quadrant, $\tan\theta$ is positive and increases from 0 to ∞; and in the fourth quadrant, $\tan\theta$ is negative and increases from $-\infty$ to 0. The cycle then repeats itself and we can draw the graph of $f(\theta) = \tan\theta$

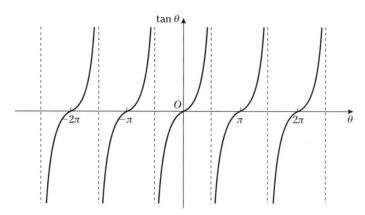

From the graph we can see that the properties of the tangent function are different from those of the sine and cosine functions.

It is not continuous, being *undefined* when $\theta = \ldots -\frac{1}{2}\pi,\ \frac{1}{2}\pi,\ \frac{3}{2}\pi,$

The range of values of tan θ is unlimited.

It is periodic with a period of π (not 2π as in the other cases).

The graph has rotational symmetry about $\theta = 0$, so

$\tan(-\alpha) = -\tan\alpha$

The graph has rotational symmetry about $\theta = \frac{1}{2}\pi$, giving

$\tan(\pi - \alpha) = -\tan\alpha$

The cycle repeats itself from $\theta = \pi$ to 2π, so

$\tan(\pi + \alpha) = \tan\alpha$ and $\tan(2\pi - \alpha) = -\tan\alpha$

Example 9c

Express $\tan\frac{11}{4}\pi$ as the tangent of an acute angle.

$$\tan\left(\frac{11}{4}\pi\right) = \tan\left(2\pi + \frac{3}{4}\pi\right) = \tan\left(\frac{3}{4}\pi\right)$$

$$= \tan\left(\pi - \frac{1}{4}\pi\right)$$

$$= -\tan\frac{1}{4}\pi$$

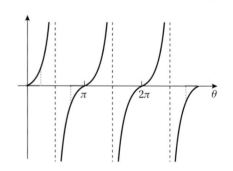

Exercise 9c

1 Find the exact value of

  (a) $\tan\frac{9}{4}\pi$ (b) $\tan 120°$

  (c) $\tan -\frac{2}{3}\pi$ (d) $\tan\frac{7}{4}\pi$

2 Write in terms of the tangent of an acute angle

  (a) $\tan 220°$ (b) $\tan\frac{12}{7}\pi$

  (c) $\tan 310°$ (d) $\tan -\frac{7}{5}\pi$

3 Sketch the graph of $y = \tan \theta$ for values of θ in the range 0 to 2π. From this sketch find the values of θ in this range for which

(a) $\tan \theta = 1$ (b) $\tan \theta = -1$

(c) $\tan \theta = 0$ (d) $\tan \theta = \infty$

4 Sketch, for $0 \leqslant \theta \leqslant 2\pi$, the graphs of

(a) $\tan 2\theta$ (b) $-\tan \theta$ (c) $1 + \tan \theta$

5 Using the basic definitions of $\sin \theta$, $\cos \theta$ and $\tan \theta$, show that, for all values of θ,

$$\tan \theta = \frac{\sin \theta}{\cos \theta}$$

RELATIONSHIPS BETWEEN $\sin \theta$, $\cos \theta$ AND $\tan \theta$

Because each trigonometric function is a ratio of two of the three quantities x, y and r, we expect to find several relationships between $\sin \theta$, $\cos \theta$ and $\tan \theta$. Here is a summary of the results so far.

If the graph of $\cos \theta$ is shifted by $\frac{1}{2}\pi$ to the right we get the graph of $\sin \theta$.

So $\cos \left(\theta - \frac{1}{2}\pi \right) = \sin \theta$

But $\cos \left(\theta - \frac{1}{2}\pi \right) = \cos \left(\frac{1}{2}\pi - \theta \right)$

Therefore $\cos \left(\frac{1}{2}\pi - \theta \right) = \sin \theta$

Two angles that add up to $\frac{1}{2}\pi$ (i.e. $90°$) are called *complementary* angles.

i.e.   **the sine of an angle is equal to the cosine of the complementary angle and vice-versa.**

Now $\sin \theta = \dfrac{y}{r}$, $\cos \theta = \dfrac{x}{r}$ and $\tan \theta = \dfrac{y}{x}$

$\therefore$ $\dfrac{\sin \theta}{\cos \theta} = \dfrac{\dfrac{y}{r}}{\dfrac{x}{r}} = \dfrac{y}{x} = \tan \theta$

i.e.   for all values of θ $\tan \theta \equiv \dfrac{\sin \theta}{\cos \theta}$

We have also seen that the sign of each trigonometric ratio depends on the size of the angle, i.e. the quadrant in which P is. We can summarise the sign of each ratio in a quadrant diagram.

$\sin + ve$	All $+ ve$
$\tan + ve$	$\cos + ve$

Examples 9d

1 Give all the values of x between 0 and $360°$ for which $\sin x = -0.3$

The value given for x by a calculator is $-17.5°$.

From the graph, we see that, when $\sin x = -0.3$, the values of x in the specified range are $180° + 17.5°$ and $360° - 17.5°$.

When $\sin x = -0.3$, $x = 197.5°$ and $342.5°$

When the range of values is given in degrees, the answer should also be given in degrees and the same applies for radians.

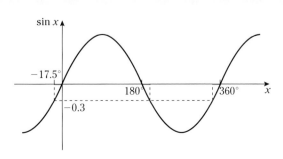

2 Find the smallest positive value of θ for which $\cos\theta = 0.7$ and $\tan\theta$ is negative.

If $\cos\theta = 0.7$, the possible values of θ are 45.6°, 314.4°, ...

Now $\tan\theta$ is positive if θ is in the first quadrant and negative if θ is in the fourth quadrant.

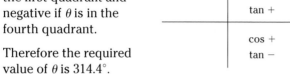

Therefore the required value of θ is 314.4°.

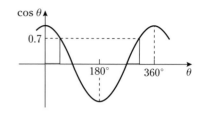

3 The function $f(x)$ is such that $f(x) = 1 - 2\sin x$ for $0 \leqslant x \leqslant 2\pi$

(a) Sketch the graph of $f(x)$ and use your sketch to give the maximum and minimum values of $f(x)$.

(b) Solve the equation $f(x) = 0$

(a)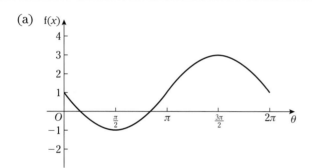

Start with a sketch of $y = 2\sin x$:

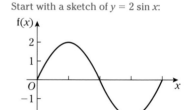

Then reflect this curve in the x-axis to give the graph of $y = -2\sin x$ and move it up the y-axis by 1 unit.

The maximum value of $f(x)$ is 3 and the minimum value is -1.

(b) $1 - 2\sin x = 0$

$\Rightarrow \quad \sin x = \dfrac{1}{2}$

$\therefore \qquad x = \dfrac{\pi}{6}, \dfrac{5\pi}{6}$

4 $f(x)$ is defined by $f(x) = 2 + a\cos(2x - \pi)$ for $0 \leqslant x \leqslant \pi$

The maximum value of $f(x)$ is 5.

(a) Find the value of a and state the minimum value of $f(x)$.

(b) Sketch the graph of $f(x)$.

(a) The maximum value of $\cos(2x - \pi)$ is 1, and so the maximum value of $a\cos(2x - \pi)$ is a

Therefore the maximum value of $f(x)$ is $2 + a$,

$\therefore \qquad 2 + a = 5 \Rightarrow a = 3$

The minimum value of $2 + a\cos(2x - \pi)$ is $2 + a(-1) = 2 - a$

$\therefore \qquad$ minimum value of $f(x)$ is -1.

(b)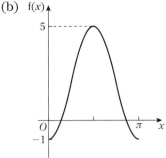

Start with a sketch of
$y = \cos(2x - \pi)$

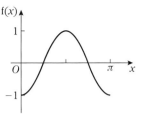

Stretch this curve by 3 units
parallel to the y-axis to give
the graph of $y = 3\cos(2x - \pi)$

Move the second curve
2 units up parallel to
the y-axis.

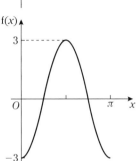

Exercise 9d

1 Within the range $-2\pi \leqslant \theta \leqslant 2\pi$, give all the values of θ for which

(a) $\sin\theta = 0.4$ (b) $\cos\theta = -0.5$

(c) $\tan\theta = 1.2$

2 Within the range $0 \leqslant \theta \leqslant 720°$, give all the values of θ for which

(a) $\tan\theta = -0.8$ (b) $\sin\theta = -0.2$

(c) $\cos\theta = 0.1$

3 Find the smallest angle (positive or negative) for which

(a) $\cos\theta = 0.8$ and $\sin\theta \geqslant 0$

(b) $\sin\theta = -0.6$ and $\tan\theta \leqslant 0$

(c) $\tan\theta = \sin\frac{1}{6}\pi$

4 Using $\tan\theta \equiv \dfrac{\sin\theta}{\cos\theta}$, show that the equation $\tan\theta = \sin\theta$ can be written as $\sin\theta(\cos\theta - 1) = 0$, provided that $\cos\theta \neq 0$

Hence find the values of θ between 0 and 2π for which $\tan\theta = \sin\theta$

5 Sketch the graph of $y = \sin 2\theta$. Use your sketch to help find the values of θ in the range $0 \leqslant \theta \leqslant 360°$ for which $\sin 2\theta = 0.4$

6 Sketch the graph of $y = \cos 3\theta$. Hence find the values of θ in the range $0 \leqslant \theta \leqslant 2\pi$ for which $\cos 3\theta = -1$

7 Sketch the graph of $y = 2\sin\left(x - \dfrac{\pi}{2}\right)$ for $0 \leqslant x \leqslant \pi$

Hence find the values of x for which $2\sin\left(x - \dfrac{\pi}{2}\right) = 0$ in the range $0 \leqslant x \leqslant \pi$

8 Sketch the graph of $f(x) = 5 - \cos 2x$ for $0 \leqslant x \leqslant 360°$

9 Sketch the graph of $y = 1 + 2\sin x$ for $0 \leqslant x \leqslant \pi$

State the maximum and minimum values of y.

10 Sketch the graph of $f(x) = 3 - \sin 2x$ for $0 \leqslant x \leqslant 2\pi$

11 The function $f: x \mapsto 3 + a\cos x$ for $0 \leqslant x \leqslant 2\pi$

Find the range of values of a for which the equation $f(x) = 0$ has no solution.

12 Sketch the graphs of $y = \theta$ and $y = 2\cos\theta$ for values of θ in the range $-\pi \leqslant \theta \leqslant \pi$

State the number of solutions of the equation $2\cos\theta = \theta$

Repeat this question with θ measured in degrees. Will the solutions be the same?

13 Measuring the angles in radians, sketch graphs to show, approximately, the value of θ for which

(a) $2\theta = 4\sin\theta$ (b) $\sin\theta = \theta^2$

(c) $\cos\theta = \theta - 1$

10 Trigonometry 2

After studying this chapter you should be able to

- use the identities $\dfrac{\sin \theta}{\cos \theta} \equiv \tan \theta$ and $\sin^2 \theta + \cos^2 \theta \equiv 1$
- find all the solutions of simple trigonometric equations lying in a specified interval
- use the notations $\sin^{-1}$, $\cos^{-1}$, $\tan^{-1}$ to denote the principal values of the inverse trigonometric relations.

TRIGONOMETRIC IDENTITIES

In this chapter we look at some trigonometric identities and some of their uses. Remember that an identity is the equivalence between two different forms of the same expression. One such identity is on page 89, that is

$$\tan \theta \equiv \frac{\sin \theta}{\cos \theta}$$

THE PYTHAGOREAN IDENTITY

For any angle θ,

$$\sin \theta = \frac{y}{r}, \cos \theta = \frac{x}{r} \text{ and } \tan \theta = \frac{y}{x}$$

Also, in right-handed triangle OPQ, $x^2 + y^2 = r^2$ (Pythagoras)

Therefore, $(\cos \theta)^2 + (\sin \theta)^2 = \left(\dfrac{x}{r}\right)^2 + \left(\dfrac{y}{r}\right)^2 = \dfrac{x^2 + y^2}{r^2} = 1$

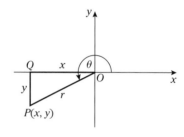

Using the notation $\cos^2 \theta$ to mean $(\cos \theta)^2$, etc., we have

$$\cos^2 \theta + \sin^2 \theta \equiv 1$$

These identities can be used to

- simplify trigonometric expressions
- eliminate trigonometric terms from pairs of equations
- derive a variety of further trigonometric relationships
- calculate other trigonometric ratios of any angle for which one trig ratio is known
- solve some trigonometric equations.

Examples 10a

1 Simplify $\dfrac{\sin \theta}{1 - \cos^2 \theta}$

$$\frac{\sin \theta}{1 - \cos^2 \theta} \equiv \frac{\sin \theta}{\sin^2 \theta}$$

$$\equiv \frac{1}{\sin \theta} \qquad \qquad \text{Using } \sin^2 \theta + \cos^2 \theta \equiv 1$$

2 Eliminate θ from the equations $x = 2 \cos \theta$ and $y = 3 \sin \theta$

$\cos\theta = \dfrac{x}{2}$ and $\sin\theta = \dfrac{y}{3}$

Using $\cos^2\theta + \sin^2\theta \equiv 1$ gives

$$\left(\frac{x}{2}\right)^2 + \left(\frac{y}{3}\right)^2 = 1$$

$$\Rightarrow \qquad 9x^2 + 4y^2 = 36$$

In Example 2, both x and y initially depend on θ, a variable angle. Used in this way, θ is called a *parameter*, and is a type of variable that plays an important part in the analysis of curves and functions.

3 If $\sin A = -\dfrac{1}{3}$ and A is in the third quadrant, find $\cos A$ without using a calculator.

There are two ways of doing this problem. The first method involves drawing a quadrant diagram and working out the remaining side of the triangle, using Pythagoras' theorem.

From the diagram, $x = -2\sqrt{2}$

$$\therefore \qquad \cos A = \frac{x}{r} = -\frac{2\sqrt{2}}{3}$$

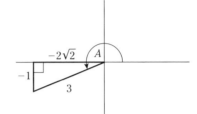

The second method uses the identity $\cos^2 A + \sin^2 A \equiv 1$ giving

$$\cos^2 A + \frac{1}{9} = 1$$

$$\Rightarrow \qquad \cos A = \pm\sqrt{\frac{8}{9}} = \pm\frac{2\sqrt{2}}{3}$$

As A is between π and $\frac{3}{2}\pi$, $\cos A$ is negative, i.e.

$$\cos A = -\frac{2\sqrt{2}}{3}$$

4 Prove that $(1 - \cos A)\left(1 + \dfrac{1}{\cos A}\right) \equiv \sin A \tan A$

Because the relationship has yet to be proved, we must not assume it is true by using the complete identity in our working. The left- and right-hand sides must be isolated throughout the proof, preferably by working on only one of these sides. Start with the more complicated side. It often helps to express all ratios in terms of sine and/or cosine as, usually, these are easier to work with.

Consider the left-hand side:

$$(1 - \cos A)\left(1 + \frac{1}{\cos A}\right) \equiv 1 + \left(\frac{1}{\cos A}\right) - \cos A - \frac{\cos A}{\cos A}$$

$$\equiv \frac{1}{\cos A} - \cos A$$

$$\equiv \frac{1 - \cos^2 A}{\cos A} \equiv \frac{\sin^2 A}{\cos A} \qquad \cos^2 A + \sin^2 A \equiv 1$$

$$\equiv \sin A\left[\frac{\sin A}{\cos A}\right] \equiv \sin A \tan A \equiv \text{right-hand side}$$

Exercise 10a

1 Without using a calculator, complete the following table.

	sin θ	cos θ	tan θ	type of angle
(a)		$-\frac{5}{13}$		reflex
(b)	$\frac{3}{5}$			obtuse
(c)			$\frac{7}{24}$	acute
(d)				straight line

Simplify the following expressions.

2 $\dfrac{\sin^2 A(1 - \cos^2 A)}{\cos^2 A(1 - \sin^2 A)}$

3 $\dfrac{\sin \theta}{\sqrt{(1 - \cos^2 \theta)}}$

4 $\dfrac{\sin \theta}{\cos \theta} + \dfrac{\cos \theta}{\sin \theta}$

5 $\dfrac{\sqrt{(1 + \tan^2 \theta)}}{\sqrt{(1 - \sin^2 \theta)}}$

6 $\dfrac{\tan \theta}{\cos \theta \sqrt{1 + \tan^2 \theta}}$

7 $\dfrac{\sin \theta \tan^2 \theta}{1 + \tan^2 \theta}$

Eliminate θ from the following pairs of equations.

8 $x = 4 \cos \theta$
   $y = 4 \sin \theta$

9 $x = \dfrac{a}{\sin \theta}$
   $y = \dfrac{b}{\cos \theta}$

10 $x = 1 + \sin \theta$
   $y = \cos \theta$

11 $x = 1 - \sin \theta$
   $y = 1 + \cos \theta$

12 $x = \dfrac{a}{\cos \theta}$
   $y = b \sin \theta$

13 $x = 2 + \tan \theta$
   $y = 2 \cos \theta$

Prove the following identities.

14 $\dfrac{1 + \tan^2 \theta}{\tan \theta} \equiv \dfrac{1}{\cos \theta \sin \theta}$

15 $\dfrac{\cos^2 A - \sin^2 A}{\cos A - \sin A} \equiv \sin A + \cos A$

16 $\dfrac{1 - \cos^2 \theta}{\sin \theta} \equiv \sin \theta$

17 $\dfrac{\sin A}{1 + \cos A} \equiv \dfrac{1 - \cos A}{\sin A}$

   (*Hint* Multiply top and bottom of left-hand side by $(1 - \cos A)$.)

18 $\dfrac{\sin A}{1 + \cos A} + \dfrac{1 + \cos A}{\sin A} \equiv \dfrac{2}{\sin A}$

SOLVING EQUATIONS

We can now solve more equations by using the Pythagorean identities.

Examples 10b

1 Solve the equation $2 \cos^2 \theta - \sin \theta = 1$ for values of θ in the range 0 to 2π.

The given equation is quadratic, but it involves the sine and the cosine of θ, so we use $\cos^2 \theta + \sin^2 \theta \equiv 1$ to express the equation in terms of sin θ only.

$2 \cos^2 \theta - \sin \theta = 1$

$\Rightarrow \qquad 2(1 - \sin^2 \theta) - \sin \theta = 1$

$\Rightarrow \qquad 2 \sin^2 \theta + \sin \theta - 1 = 0$

$\Rightarrow \qquad (2 \sin \theta - 1)(\sin \theta + 1) = 0$

$\Rightarrow \qquad \sin \theta = \frac{1}{2} \quad \text{or} \quad -1$

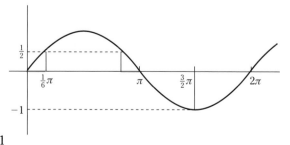

If $\sin \theta = \frac{1}{2}$, $\theta = \frac{1}{6}\pi$, $\frac{5}{6}\pi$

If $\sin \theta = -1$, $\theta = \frac{3}{2}\pi$

Therefore the solution of the equation is $\theta = \frac{1}{6}\pi$, $\frac{5}{6}\pi$, $\frac{3}{2}\pi$

2 (a) Explain why $\cos x = 0$ is not a possible solution of the equation $1 = \sin x \tan x$

(b) Solve the equation $1 = \sin x \tan x$ for values of x from 0 to 360°.

(a) Using $\tan x = \dfrac{\sin x}{\cos x}$

$$1 = \sin x \tan x \quad \Rightarrow \quad 1 = \frac{\sin^2 x}{\cos x}$$

Both sides of an equation can be multiplied by any number *except* zero. We can simplify the equation by multiplying by $\cos x$ provided that $\cos x \neq 0$. So we must exclude any values of x for which $\cos x = 0$ from the solutions.

(b) $1 = \dfrac{\sin^2 x}{\cos x} \quad \Rightarrow \quad \cos x = \sin^2 x$

$\Rightarrow \quad \cos^2 x + \cos x - 1 = 0$

This equation is a quadratic equation and does not factorise, so we use the formula, giving

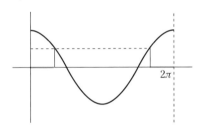

$\cos x = \frac{1}{2}\left(-1 \pm \sqrt{5}\right)$

$\therefore \qquad \cos x = 0.618\ldots$ Do not correct this value, leave it in your calculator.

$\Rightarrow \qquad x = 51.8° \quad \text{or} \quad 308.2°$ $\cos x \neq 0$ for either value of x

or $\cos x = -1.618$ and there is no value of x for which this is true.

Exercise 10b

Solve the following equations for angles in the range $0 \leqslant \theta \leqslant 360°$

1 $\sin \theta = \dfrac{\sqrt{3}}{2}$

2 $\cos \theta = 0$

3 $\tan \theta = -\sqrt{3}$

4 $\sin \theta = -\dfrac{1}{4}$

5 $\cos \theta = -\dfrac{1}{2}$

6 $\tan \theta = 1$

7 $\sin^2 \theta = \dfrac{1}{4}$

8 $4 \cos^2 \theta + 5 \sin \theta = 3$

9 $5 \cos \theta - 4 \sin^2 \theta = 2$

Solve the following equations for angles in the range $-\pi \leqslant \theta \leqslant \pi$

10 $2 \cos \theta - 4 \sin^2 \theta + 2 = 0$

11 $2 \sin \theta \cos \theta + \sin \theta = 0$

12 $\sqrt{3} \tan \theta = 2 \sin \theta$

EQUATIONS INVOLVING MULTIPLE ANGLES

Many trigonometric equations involve ratios of a multiple of θ, for example

$$\cos 2\theta = \tfrac{1}{2}, \ \tan 3\theta = -2$$

Note that $\cos 2\theta = \tfrac{1}{2} \ \Rightarrow \ \cos \theta = \tfrac{1}{4}$, etc.

Simple equations like this can be solved by finding first the values of the multiple angle and then, by division, the corresponding values of θ.

However, if values of θ are required in the range $\alpha \leqslant \theta \leqslant \beta$, then values of the multiple angle, $k\theta$ say, must be found in a range that is multiplied by the same factor, i.e. $k\alpha \leqslant k\theta \leqslant k\beta$

e.g.   if values of θ are required for $0 \leqslant \theta \leqslant 360°$

then 2θ must be found in the range $0 \leqslant 2\theta \leqslant 720°$

$\tfrac{1}{2}\theta$ must be found in the range $0 \leqslant \tfrac{1}{2}\theta \leqslant 180°$, and so on.

Examples 10c

1 Find the values of θ in the range $-\pi \leqslant \theta \leqslant \pi$, for which $\cos 2\theta = \tfrac{1}{2}$

Using $2\theta = x$ gives $\cos x = \tfrac{1}{2}$

As values of θ are required in the range $-\pi \leqslant \theta \leqslant \pi$, we want values of x (i.e. 2θ) in the range $-2\pi \leqslant x \leqslant 2\pi$

In the range $-2\pi \leqslant x \leqslant 2\pi$ the solutions of $\cos x = \tfrac{1}{2}$ are

$\pm \tfrac{1}{3}\pi, \ \pm \tfrac{5}{3}\pi.$

But $x = 2\theta$, therefore $2\theta = \pm \tfrac{1}{3}\pi, \ \pm \tfrac{5}{3}\pi$

$\Rightarrow \qquad \theta = \pm \tfrac{1}{6}\pi, \ \pm \tfrac{5}{6}\pi$

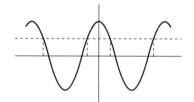

2 Find the solution of the equation $\tan\left(\tfrac{1}{3}\theta - 90°\right) = 1$ for which $0 \leqslant \theta \leqslant 540°$

Using $\quad \tfrac{1}{3}\theta - 90° = x$ gives

$\tan\left(\tfrac{1}{3}\theta - 90°\right) = \tan x$

As θ is required in the range $0 \leqslant \theta \leqslant 540°$ we must find x in the range
$\tfrac{1}{3}(0) - 90° \leqslant x \leqslant \tfrac{1}{3}(540°) - 90°$ i.e. $-90° \leqslant x \leqslant 90°$

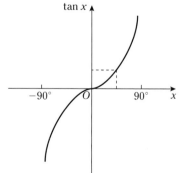

The solution of the equation $\tan x = 1$ is

$$x = 45°$$

But $\qquad\qquad\qquad x = \tfrac{1}{3}\theta - 90°$

so $\qquad\qquad\qquad \tfrac{1}{3}\theta - 90° = 45°$

$\Rightarrow \qquad\qquad\qquad \tfrac{1}{3}\theta = 135°$

i.e. $\qquad\qquad\qquad \theta = 405°$

Exercise 10c

Find the solutions of the following equations, for values of θ in the range $0 \leqslant \theta \leqslant 180°$

1 $\tan 2\theta = 1$

2 $\cos 3\theta = -0.5$

3 $\sin \frac{1}{2}\theta = -\dfrac{\sqrt{2}}{2}$

4 $\cos(2\theta - 45°) = 0$

5 $\sin\left(\frac{1}{4}\theta + 30°\right) = -1$

6 $\tan(\theta - 60°) = 0$

Solve the equations for values of θ in the range $-180° \leqslant \theta \leqslant 180°$

7 $\tan 2\theta = 1.8$

8 $\sin 3\theta = 0.7$

9 $\cos \frac{1}{2}\theta = 0.85$

Solve the equations for values of θ in the range $0 \leqslant \theta \leqslant \pi$

10 $\tan 4\theta = -\sqrt{3}$

11 $\cos\left(\theta + \frac{1}{4}\pi\right) = \frac{1}{2}$

12 $\tan\left(2\theta - \frac{1}{3}\pi\right) = -1$

THE INVERSE TRIGONOMETRIC FUNCTIONS

The function f is given by $f : x \mapsto \sin x$ for $x \in \mathbb{R}$

The inverse mapping is given by $\sin x \mapsto x$ but this is not a function because one value of $\sin x$ maps to many values of x, so $f(x) = \sin x$ does not have an inverse function for the domain $x \in \mathbb{R}$

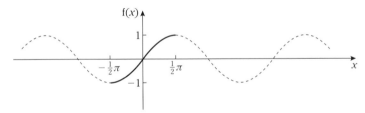

If we now look at the function $f : x \mapsto \sin x$ for $-\frac{1}{2}\pi \leqslant x \leqslant \frac{1}{2}\pi$, it is a one–one function so the inverse mapping, $\sin x \mapsto x$, is such that one value of $\sin x$ maps to only one value of x. Therefore $f : x \mapsto \sin x$ for $-\frac{1}{2}\pi \leqslant x \leqslant \frac{1}{2}\pi$ does have an inverse, so f^{-1} exists.

The equation of the graph of f is $y = \sin x$ for $-\frac{1}{2}\pi \leqslant x \leqslant \frac{1}{2}\pi$ and the curve $y = f^{-1}(x)$ is obtained by reflecting $y = \sin x$ in the line $y = x$

Therefore interchanging x and y gives the equation of this curve, i.e. $\sin y = x$, so $y =$ the angle between $-\frac{1}{2}\pi$ and $\frac{1}{2}\pi$ whose sine is x.

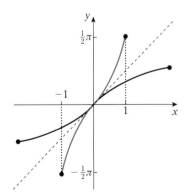

Using $\sin^{-1}$ to mean 'the angle between $-\frac{1}{2}\pi$ and $\frac{1}{2}\pi$ whose sine is', we have $y = \sin^{-1}x$

Thus   **if** $f : x \mapsto \sin x$ $-\frac{1}{2}\pi \leqslant x \leqslant \frac{1}{2}\pi$

then $f^{-1} : x \mapsto \sin^{-1}x$ $-1 \leqslant x \leqslant 1$

It is important to realise that $\sin^{-1} x$ is an angle, and that this angle is in the interval $\left[-\frac{1}{2}\pi, \frac{1}{2}\pi\right]$. The angles in this interval are called the principal values of $\sin^{-1} x$.

So, for example, $\sin^{-1} 0.5$ is the angle between $-\frac{1}{2}\pi$ and $\frac{1}{2}\pi$ whose sine is 0.5,

i.e. $\sin^{-1} 0.5 = \frac{1}{6}\pi$

Now consider the function given by $f : x \mapsto \cos x, \ 0 \leqslant x \leqslant \pi$

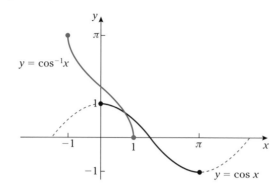

f is a one–one function so f^{-1} exists and it is denoted by $\cos^{-1}$ where $\cos^{-1} x$ means 'the angle between 0 and π whose cosine is x'.

Thus   **if** $\mathbf{f} : x \mapsto \cos x$ $0 \leqslant x \leqslant \pi$

        then $\mathbf{f^{-1}} : x \mapsto \cos^{-1} x$ $-1 \leqslant x \leqslant 1$

Note that $\cos^{-1} x$ is an angle in the interval $0 \leqslant x \leqslant \pi$. The angles in this interval are called the principal values of $\cos^{-1} x$.

So, for example, $\cos^{-1} (-0.5) = x$ $\Rightarrow$ $x = \frac{2}{3}\pi$

Similarly, if $f : x \mapsto \tan x$ for $-\frac{1}{2}\pi \leqslant x \leqslant \frac{1}{2}\pi$, then f^{-1} exists and is written $\tan^{-1}$ where $\tan^{-1} x$ means 'the angle between $-\frac{1}{2}\pi$ and $\frac{1}{2}\pi$ whose tangent is x'.

The angles between $-\frac{1}{2}\pi$ and $\frac{1}{2}\pi$ are called the principal values of $\tan^{-1} x$.

Note that the domain of $\tan^{-1} x$ is all values of x.

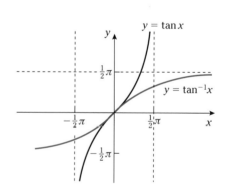

Exercise 10d

Find the value, in terms of π, of

1   $\tan^{-1} \sqrt{3}$

2   $\sin^{-1} (-1) \ \sin^{-1} \left(\frac{1}{2} \right)$

3   $\cos^{-1} 0$

4   $\sin^{-1} \left(-\frac{\sqrt{3}}{2} \right)$

5   $\cos^{-1} \left(-\frac{1}{2} \right)$

6   $\tan^{-1} (-1)$

7   $\sin^{-1} \left(\frac{1}{\sqrt{2}} \right)$

8   $\cos^{-1} \left(\frac{\sqrt{2}}{2} \right)$

9   $\tan^{-1} 1$

10   $f(x) = 4 - 2 \cos x$ for $0 \leqslant x \leqslant \pi$

   (a)   Sketch the graph of $y = f(x)$

   (b)   State whether f has an inverse.
          Give a reason for your answer.

11   $f : x \mapsto \tan \frac{1}{2} x$

   (a)   Sketch the graph of $y = f(x)$
          for $-\pi \leqslant x \leqslant 2\pi$

   (b)   Explain why, for the domain
          $-\pi \leqslant x \leqslant 2\pi$, f does not have an
          inverse.

   (c)   State the domain for which f does have
          an inverse.

Mixed exercise 10

1 Eliminate α from the equations $x = \cos \alpha$, $y = \dfrac{1}{\sin \alpha}$

2 If $\cos \beta = 0.5$, find possible values for $\sin \beta$ and $\tan \beta$, giving your answers in exact form.

3 Simplify the expression
$\dfrac{1}{1 + \cos \theta} + \dfrac{1}{1 - \cos \theta}$. Hence solve the equation $\dfrac{1}{1 + \cos \theta} + \dfrac{1}{1 - \cos \theta} = 4$ for values of θ in the range $0 \leqslant \theta \leqslant 2\pi$

4 On the same set of axes, sketch the graphs of $y = 2 \sin x$ and $y = x$ for $-\pi \leqslant x \leqslant \pi$

   Hence give the number of solutions of the equation $2 \sin x = x$

5 Solve the equation $\tan^{-1} 1 = 3\theta - 1$

6 Find the values of θ for which $\tan\left(3\theta - \tfrac{1}{3}\pi\right) = 1$ in the interval $[-\pi, \pi]$.

7 Eliminate θ from the equations $x - 2 = \sin \theta$, $y + 1 = \cos \theta$

8 $f(x) = 1 - \cos x$ for $0 \leqslant x \leqslant \theta$

   (a) State the maximum value of θ for which f^{-1} exists.

   (b) Find the value of $f^{-1}(1)$.

9 Prove that
$(\cos A + \sin A)^2 + (\cos A - \sin A)^2 \equiv 2$

10 Simplify $(1 + \cos A)(1 - \cos A)$.

11 Find the solution of the equation $\tan \theta = 3 \sin \theta$ for values of θ in the range $-180° \leqslant \theta \leqslant 180°$

12 The function $f(x)$ is given by
$f(x) = (a \sin bx) + c$ for $0 \leqslant x \leqslant \pi$

   $f(x)$ has a maximum value 8 and a minimum value of -2.

   There are four values of x for which $f(x) = 0$

   Find the values of a, b and c.

13 Solve the equation $\cos 3\theta = \tfrac{1}{2}\sqrt{3}$ giving values of θ from 0 to 180°.

14 Find, in the range $-180° \leqslant \theta \leqslant 180°$, the values of θ that satisfy the equation $2 \cos^2 \theta - \sin \theta = 1$

15 Find the solutions, in the range from 0 to π, of the equation $\tan\left(2\theta - \tfrac{1}{2}\pi\right) = \tfrac{1}{3}\sqrt{3}$

16 $f : x \mapsto 1 - 2 \sin \tfrac{1}{2}x$

   (a) Sketch the graph of $y = f(x)$ for $-2\pi \leqslant x \leqslant 2\pi$

   (b) Give a domain for f for which f^{-1} exists.

11 Sequences and series

After studying this chapter you should be able to

- recognise arithmetic and geometric progressions
- use the formulae for the nth term and for the sum of the first n terms to solve problems involving arithmetic or geometric progressions
- use the condition for the convergence of a geometric progression, and the formula for the sum to infinity of a convergent geometric progression
- use the expansion of $(a + b)^n$ where n is a positive integer.

Sequences

A sequence is an ordered set of numbers, for example 2, 4, 6, 8, 10.

There is a first term, a second term and so on.

A sequence is also called a progression.

ARITHMETIC PROGRESSIONS

Each term of the sequence 5, 8, 11, 14, 17, ... 29 is 3 more than the term before it.

So this sequence can be written as $5, \quad 5 + 3, \quad 5 + 2 \times 3, \quad 5 + 3 \times 3, \quad 5 + 4 \times 3, \quad ... 5 + 8 \times 3.$

This sequence is an example of an arithmetic progression.

An arithmetic progression (AP) is a sequence where each term differs from the next term by a constant called the common difference.

For example, the first six terms of an AP whose first term is 8 and whose common difference is -3 are

$8, 5, 2, -1, -4, -7.$

The first four terms of an AP whose first term is a and whose common difference is d are

$a, a + d, a + 2d, a + 3d.$

The nth term is $a + (n - 1)d.$

So **an AP with n terms can be written as**

$$a, (a + d), (a + 2d), ... , [a + (n - 1)d]$$

Examples 11a

1 The eighth term of an AP is 11 and the 15th term is 21. Find the common difference, the first term of the series, and the nth term.

The first term of the series is a and the common difference is d, so the eighth term is $a + 7d$,

$\therefore \quad a + 7d = 11$ $\qquad\qquad\qquad$ [1]

and the 15th term is $a + 14d$,

$\therefore \quad a + 14d = 21$ $\qquad\qquad\qquad$ [2]

$[2] - [1]$ gives $7d = 10 \quad \Rightarrow \quad d = \frac{10}{7}$ and $a = 1$

so the first term is 1 and the common difference is $\frac{10}{7}$.

Hence the nth term is $a + (n - 1)d = 1 + (n - 1)\frac{10}{7} = \frac{1}{7}(10n - 3)$

2 The nth term of an AP is $12 - 4n$. Find the first term and the common difference.

The nth term is $12 - 4n$, so the first term ($n = 1$) is 8.

The second term ($n = 2$) is 4.

Therefore the common difference is -4.

THE SUM OF AN ARITHMETIC PROGRESSION

The first ten even numbers is an AP.

Writing the sum of the AP first in normal, then in reverse, order gives

$$S = \ \ 2 + \ \ 4 + \ \ 6 + \ \ 8 + \dots + 18 + 20$$
$$S = 20 + 18 + 16 + 14 + \dots + \ \ 4 + \ \ 2$$

Adding gives $2S = 22 + 22 + 22 + 22 + \dots + 22 + 22$

As there are ten terms in this series, so

$$2S = 10 \times 22 \quad \Rightarrow \quad S = 110$$

Applying this method to a general AP gives formulae for the sum, which can be quoted and used.

S_n is the sum of the first n terms of an AP with last term l,

Therefore $S_n = \quad a \quad + (a + d) + (a + 2d) + \ \dots \ + (l - d) + \quad l$

reversing $S_n = \quad l \quad + (l - d) + (l - 2d) + \ \dots \ + (a + d) + \quad a$

adding $2S_n = (a + l) + (a + l) + \ (a + l) \ + \ \dots \ + (a + l) + (a + l)$

as there are n terms, $2S_n = n(a + l)$

$\Rightarrow$ $$\boldsymbol{S_n = \tfrac{1}{2}n\,(a + l)}$$

i.e. S_n = (number of terms) $\times$ (half the sum of the first and last terms)

Also, because the nth term, l, is equal to $a + (n - 1)d$,

$$S_n = \tfrac{1}{2}n[a + a + (n - 1)d]$$

i.e $$\boldsymbol{S_n = \tfrac{1}{2}n[2a + (n - 1)d]}$$

Either of these formulae can now be used to find the sum of the first n terms of an AP.

Examples 11a cont.

3 Find the sum of the following sequences.

(a) an AP of 11 terms whose first term is 1 and whose last term is 6

(b) $\frac{4}{3}, \frac{2}{3}, 0, -\frac{2}{3}, \dots -\frac{10}{3}$

(a)   We know the first and last terms, and the number of terms, so use $S_n = \frac{1}{2}n(a + l)$

$\Rightarrow \quad S_{11} = \frac{11}{2}(1 + 6) = \frac{77}{2}$

(b) $\frac{4}{3}, \frac{2}{3}, 0, -\frac{2}{3}, \dots -\frac{10}{3}$

This is an AP where $a = \frac{4}{3}$, $d = -\frac{2}{3}$

$-\frac{10}{3} = \frac{4}{3} + (n - 1) \times -\frac{2}{3} \quad \Rightarrow \quad n = 8$ Using nth term $= a + (n - 1)d$

$S_8 = \frac{1}{2} \times 8 \times \left(\frac{4}{3} - \frac{10}{3}\right) = -8$

4 The sum of the first ten terms of an AP is 50 and the fifth term is three times the second term. Find the first term and the sum of the first 20 terms.

a is the first term and d is the common difference, and there are n terms. Using $S = \frac{1}{2}n[2a + (n - 1)d]$ gives

$$S_{10} = 50 = 5(2a + 9d) \qquad\qquad [1]$$

Using $\qquad\qquad u_n = a + (n - 1)d$ gives

$$u_5 = a + 4d \quad \text{and} \quad u_2 = a + d \text{ where } u_n \text{ is the } n\text{th term}$$

Therefore $\qquad a + 4d = 3(a + d) \qquad\qquad [2]$

From [1] and [2]: $d = 1$ and $a = \frac{1}{2}$

So the first term is $\frac{1}{2}$ and the sum of the first 20 terms is S_{20} where

$$S_{20} = 10(1 + 19 \times 1) = 200$$

5 The sum of the first n terms of a progression is given by $S_n = n(n + 3)$

Find the fourth term of the progression and show that the terms are in arithmetic progression.

If the terms of the progression are $a_1, a_2, a_3, \dots, a_n$

then $\qquad\qquad S_n = a_1 + a_2 + \dots + a_n = n(n + 3)$

So $\qquad\qquad S_4 = a_1 + a_2 + a_3 + a_4 = 28$

and $\qquad\qquad S_3 = a_1 + a_2 + a_3 = 18$

$\qquad\qquad S_4 - S_3 = a_4 = 28 - 18$

Hence the fourth term, a_4, is 10.

Now $\qquad\qquad S_n = a_1 + a_2 + \dots + a_{n-1} + a_n = n(n + 3)$

and $\qquad\qquad S_{n-1} = a_1 + a_2 + \dots + a_{n-1} = (n - 1)(n + 2)$ replacing n with $n - 1$

Hence the nth term, a_n, is given by $S_n - S_{n-1}$

$$a_n = n(n + 3) - (n - 1)(n + 2) = 2n + 2$$

Replacing n by $n - 1$ gives the $(n - 1)$th term

i.e. $\qquad\qquad a_{n-1} = 2(n - 1) + 2 = 2n$

Then $\qquad a_n - a_{n-1} = (2n + 2) - 2n = 2$

i.e. there is a common difference of 2 between successive terms, showing that the progression is an AP.

Exercise 11a

1 Write down the fifth term and the nth term of the following APs.

   (a) first term 5, common difference 3

   (b) first term 6, common difference -2

   (c) first term p, common difference q

   (d) first term 10, last term 30, 11 terms

   (e) $1, 5, \ldots$

   (f) $2, 1\frac{1}{2}, \ldots$

   (g) $-4, -1, \ldots$

2 Find the sum of the first ten terms of each AP given in question 1.

3 The ninth term of an AP is 8 and the fourth term is 20. Find the first term and the common difference.

4 The sixth term of an AP is twice the third term and the first term is 3. Find the common difference and the tenth term.

5 The nth term of an AP is $\frac{1}{2}(3 - n)$.

   Write down the first three terms and the 20th term.

6 Find the sum, to the number of terms given, of each of the following APs.

   (a) $1 + 2\frac{1}{2} + \ldots$, six terms

   (b) $3 + 5 + \ldots$, eight terms

   (c) $a_1 + a_2 + a_3 + \ldots + a_8$ where $a_n = 2n + 1$

   (d) $4 + 6 + 8 + \ldots + 20$

   (e) $S_n = n^2 - 3n$, eight terms

7 The sum of the first n terms of an AP is S_n where $S_n = n^2 - 3n$. Write down the fourth term and the nth term.

8 The sum of the first n terms of a progression is given by S_n where $S_n = n(3n - 4)$. Show that the progression is an AP.

9 In an arithmetic progression, the eighth term is twice the fourth term and the 20th term is 20. Find the common difference and the sum of the terms from the eighth to the 20th inclusive.

10 The sum of the first n terms of a progression is S_n where $S_n = 2n^2 - n$

   Prove that the progression is an AP, stating the first term and the common difference.

GEOMETRIC PROGRESSIONS

The sequence

$$12, \ 6, \ 3, \ 1.5, \ 0.75, \ 0.375, \ \ldots$$

is such that each term is half the preceding term. The sequence can be written

$$12, \ 12\left(\tfrac{1}{2}\right), \ 12\left(\tfrac{1}{2}\right)^2, \ 12\left(\tfrac{1}{2}\right)^3, \ 12\left(\tfrac{1}{2}\right)^4, \ 12\left(\tfrac{1}{2}\right)^5, \ \ldots$$

This sequence is an example of a geometric progression (GP). A geometric progression is one where each term is a constant multiple of the term immediately before it. This constant multiple is called the common ratio and it can have any value.

Hence, if a GP has a first term of 3 and a common ratio of -2, the first four terms are

$$3, \ 3(-2), \ 3(-2)^2, \ 3(-2)^3$$

i.e. $\quad 3, \ -6, \ 12, \ -24$

When a GP has a first term a, and a common ratio r, the first four terms are a, ar, ar^2, ar^3 and the nth is ar^{n-1},

i.e. **a GP with n terms can be written $a, ar, ar^2, \ldots, ar^{n-1}$**

THE SUM OF THE FIRST n TERMS OF A GEOMETRIC PROGRESSION

The sum of the first eight terms, S_8, of the GP with first term 1 and common ratio 3 is given by

$$S_8 = 1 + 1(3) + 1(3)^2 + 1(3)^3 + \ldots + 1(3)^7$$

$$\Rightarrow \qquad 3S_8 = 3 + 3^2 + 3^3 + \ldots + 3^7 + 3^8$$

Hence $\quad S_8 - 3S_8 = 1 + 0 + 0 + 0 + \ldots + 0 - 3^8$

So $\qquad S_8(1 - 3) = 1 - 3^8$

$$\Rightarrow \qquad S_8 = \frac{1 - 3^8}{1 - 3} = \frac{3^8 - 1}{2}$$

This process can be applied to any GP.

The sum, S_n, of the first n terms of a GP with first term a and common ratio r

is given by $\qquad S_n = a + ar + \ldots + ar^{n-2} + ar^{n-1}$

Multiplying by r gives $\qquad rS_n = ar + ar^2 + \ldots + ar^{n-1} + ar^n$

Hence $\quad S_n - rS_n = a - ar^n$

$$\Rightarrow \qquad S_n(1 - r) = a(1 - r^n)$$

$$\Rightarrow \qquad \boldsymbol{S_n = \frac{a(1 - r^n)}{1 - r}}$$

If $\quad r > 1$ the formula can be written $\dfrac{a(r^n - 1)}{r - 1}$.

Examples 11b

1 The fifth term of a GP is 8, the third term is 4, and the common ratio is positive.
   Find the first term, the common ratio, and the sum of the first ten terms.

   For a first term a and common ratio r, the nth term is ar^{n-1}

   Therefore $\qquad ar^4 = 8 \qquad (n = 5)$

   and $\qquad ar^2 = 4 \qquad (n = 3)$

   dividing gives $\qquad r^2 = 2$

   $\Rightarrow \qquad r = \sqrt{2} \quad$ and $\quad a = 2$

   Using the formula $\quad S_n = \dfrac{a(r^n - 1)}{r - 1}$ gives

   $$S_{10} = \frac{2\left[(\sqrt{2})^{10} - 1\right]}{\sqrt{2} - 1} = \frac{62}{\sqrt{2} - 1} = 150 \ (3 \text{ s.f.})$$

2 The sum of the first n terms of a series is $3^n - 1$. Show that the terms of this series are in
   geometric progression and find the first term and the common ratio.

If the series is $\qquad a_1 + a_2 + \ldots + a_n$

then $\qquad S_n = a_1 + a_2 + \ldots + a_{n-1} + a_n = 3^n - 1$

and $\qquad S_{n-1} = a_1 + a_2 + \ldots \qquad + a_{n-1} = 3^{n-1} - 1$

therefore $S_n - S_{n-1}$ gives $\qquad\qquad a_n = 3^n - 1 - (3^{n-1} - 1)$

i.e. $\quad$ the nth term is $\quad 3^n - 3^{n-1} = 3^{n-1}(3 - 1) = (2)3^{n-1}$

The first term is given by $n = 1 \quad \Rightarrow \quad a = 2(3^0) = 2$

Similarly $a_{n-1} = (2)3^{n-2} \quad$ so $\quad a_n \div a_{n-1} = 3$

showing that successive terms in the series have a constant ratio of 3.
Hence this series is a GP with first term 2 and common ratio 3.

Exercise 11b

1 Write down the fifth term of the following GPs.

(a) $2, 4, 8, \ldots$

(b) $2, 1, \frac{1}{2}, \ldots$

(c) $3, -6, 12, \ldots$

(d) first term 8, common ratio $-\frac{1}{2}$

(e) first term 3, last term $\frac{1}{81}$, six terms

2 Find the sum, to the number of terms given, of the following GPs.

(a) $3 + 6 + \ldots$, six terms

(b) $3 - 6 + \ldots$, eight terms

(c) $1 + \frac{1}{2} + \frac{1}{4} + \ldots$, 20 terms

(d) first term 5, common ratio $\frac{1}{5}$, five terms

(e) first term $\frac{1}{2}$, common ratio $-\frac{1}{2}$, ten terms

(f) first term 1, common ratio -1, 2001 terms

3 The sixth term of a GP is 16 and the third term is 2. Find the first term and the common ratio.

4 Find the common ratio, given that it is negative, of a GP whose first term is 8 and whose fifth term is $\frac{1}{2}$.

5 The nth term of a GP is $\left(-\frac{1}{2}\right)^n$
Write down the first term and the tenth term.

6 Find the sum to n terms of the following progressions.

(a) $x, x^2, x^3, \ldots$ $\qquad$ (b) $x, 1, \frac{1}{x}, \ldots$

(c) $1, y, y^2, \ldots$

7 Find the sum of the first n terms of the GP $2 + \frac{1}{2} + \frac{1}{8} + \ldots$

8 The sum of the first three terms of a GP is 14. If the first term is 2, find the sum of the first five terms, given that the common ratio is positive.

THE SUM TO INFINITY OF A GEOMETRIC PROGRESSION

Any GP can be written as $a, ar, ar^2, \ldots$

The sum of the first n terms, S_n, is given by $S_n = \dfrac{a(1 - r^n)}{1 - r}$

When $-1 < r < 1$, as n gets larger, r^n gets closer to zero, so $\lim\limits_{n \to \infty} (r^n) = 0$

So $\quad \lim\limits_{n \to \infty} S_n = \lim\limits_{n \to \infty} \left[\dfrac{a(1 - r^n)}{1 - r}\right] = \dfrac{a}{1 - r}$

This is not true when $r > 1$ or $r < -1$, because as n gets larger r^n does not get closer to zero.

$\lim\limits_{n \to \infty} S_n$ is called the sum to infinity and is denoted by S_∞

$$\text{for a GP,} \quad S_\infty = \frac{a}{1 - r} \text{ provided that } |r| < 1$$

Examples 11c

1 Find the sum to infinity of the GP $1, -\frac{1}{4}, +\frac{1}{16}, -\frac{1}{64}, + \ldots$

$$1, -\tfrac{1}{4}, +\tfrac{1}{16}, -\tfrac{1}{64}, + \ldots = 1, + \left(-\tfrac{1}{4}\right), + \left(-\tfrac{1}{4}\right)^2, + \left(-\tfrac{1}{4}\right)^3, + \ldots$$

$\therefore \quad a = 1 \quad$ and $\quad r = -\tfrac{1}{4}$

$\therefore \quad S_{\infty} = \dfrac{a}{1 - r} = \dfrac{1}{1 - \left(-\tfrac{1}{4}\right)} = \dfrac{4}{5}$

2 The first term of a GP is 1. The third term is the mean of the first and second terms. The GP has a sum to infinity. Find the common ratio and the sum to infinity.

$a = 1$ so the GP is $1, r, r^2, r^3, \ldots$

$\therefore \qquad\qquad r^2 = \tfrac{1}{2}(1 + r)$

$\Rightarrow \qquad 2r^2 - r - 1 = 0$

$\therefore \quad (2r + 1)(r - 1) = 0$

The GP has a sum to infinity, $\quad \therefore r = -\tfrac{1}{2}$

$S_{\infty} = \dfrac{a}{1 - r} = \dfrac{1}{1 - \left(-\tfrac{1}{2}\right)} = \dfrac{2}{3}$

Exercise 11c

1 Find the sum to infinity of the progressions.

(a) $4, \dfrac{4}{3}, \dfrac{4}{3^2}, \ldots$

(b) $20, -10, +5, -2.5, + \ldots$

(c) $\dfrac{5}{10}, \dfrac{5}{100}, \dfrac{5}{1000}, \ldots$

(d) $3, -1, +\dfrac{1}{3}, -\dfrac{1}{9}, + \ldots$

2 The sum to infinity of a GP is twice the first term. Find the common ratio.

3 The sum to infinity of a GP is 16 and the sum of the first four terms is 15. Find the first four terms.

4 Find the sum to infinity of the GP
$1, -\tfrac{1}{2}, \tfrac{1}{4}, -\tfrac{1}{8}, \ldots$

The remaining questions involve APs and GPs.

5 Find the sum of the progression
$2, -(2)(3), (2)(3)^2, -(2)(3)^3, \ldots, (2)(3)^{10}$.

6 The fourth term of an AP is 8 and the sum of the first ten terms is 40.

Find the first term and the tenth term.

7 The second, fourth and eighth terms of an AP are the first three terms of a GP.

Find the common ratio of the GP.

8 Find the value of x for which the numbers $x + 1, x + 3, x + 7$ are in geometric progression.

9 The second term of a GP is $\tfrac{1}{2}$ and the sum to infinity is 4.

Find the first term and the common ratio of the GP.

Series

A series is formed when the terms of a progression are added together.

For example, $\quad 1, 5, 9, 13, \ldots$ is a progression,

and $\quad 1 + 5 + 9 + 13 + \ldots$ is a series.

We now look at the series formed when $(a + b)^n$ is expanded, where n is a positive integer. The sum of two terms, such as $a + b$ is called a binomial.

Pascal's Triangle

We can expand, for example, $(a + b)^5$ by multiplying out the brackets, but a quicker method is to use Pascal's Triangle.

First look at these expansions:

$(a + b)^1 = a + b$

$(a + b)^2 = a^2 + 2ab + b^2$

$(a + b)^3 = a^3 + 3a^2b + 3ab^2 + b^2$

$(a + b)^4 = a^4 + 4a^3b + 6a^2b^2 + 4ab^3 + b^4$

Notice that the powers of a and b form a pattern.

From the expansion of $(a + b)^4$ you can see that the first term is a^4 and then the power of a decreases by 1 in each succeeding term while the power of b increases by 1. In all the terms, the sum of the powers of a and b is 4. There is a similar pattern in the other expansions.

Now look at just the coefficients of the terms. Writing these in a triangular array gives

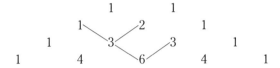

This array is called Pascal's Triangle and it also has a pattern:
each row starts and ends with 1 and each other number is the sum of the two numbers in the row above it as shown.

Now you can write down as many rows as you need.

For example to expand $(a + b)^6$, go as far as row 6:

$$\begin{array}{ccccccccccccc}
& & & & & & 1 & & 1 & & & & \\
& & & & & 1 & & 2 & & 1 & & & \\
& & & & 1 & & 3 & & 3 & & 1 & & \\
& & & 1 & & 4 & & 6 & & 4 & & 1 & \\
& & 1 & & 5 & & 10 & & 10 & & 5 & & 1 \\
& 1 & & 6 & & 15 & & 20 & & 15 & & 6 & & 1
\end{array}$$

Using what we know about the pattern of the powers and using row six of the array gives

$(a + b)^6 = a^6 + 6a^5b + 15a^4b^2 + 20a^3b^3 + 15a^2b^4 + 6ab^5 + b^6$

THE BINOMIAL THEOREM

We could use Pascal's Triangle to expand $(a + b)^n$ for any positive integer n, but this can take lots of time.

The binomial theorem gives a method for expanding $(a + b)^n$.

The binomial theorem states that when n is a positive integer,

$$(a + b)^n = a^n + na^{n-1}b + \frac{n(n-1)}{2 \times 1}\, a^{n-2}b^2 + \frac{n(n-1)(n-2)}{3 \times 2 \times 1}\, a^{n-3}b^3 + \ldots + b^n$$

The right-hand side of this identity is called the series expansion of $(a + b)^n$. The coefficients of the powers of a and b are called binomial coefficients. The denominators of these coefficients involve the product of all the positive integers from the power of b down to 1. We can write these using factorial notation.

The product $4 \times 3 \times 2 \times 1$ is called 4 factorial and is written 4!
8! means $8 \times 7 \times 6 \times 5 \times 4 \times 3 \times 2 \times 1$

The binomial coefficient $\dfrac{n(n-1)(n-2)}{3 \times 2 \times 1}$ can be written as $\dfrac{n(n-1)(n-2)}{3!}$
or even more concisely as $\binom{n}{3}$.

So $\binom{n}{6}$ means $\dfrac{n(n-1)(n-2)(n-3)(n-4)(n-5)}{6!}$

Therefore we can write the expansion of $(a+b)^n$ as

$$(a+b)^n = a^n + na^{n-1}b + \binom{n}{2}a^{n-2}b^2 + \binom{n}{3}a^{n-3}b^3 + \binom{n}{4}a^{n-4}b^4 + \ldots + b^n$$

Notice that

1 the expansion of $(a+b)^n$ is a finite series with $n+1$ terms,

2 the coefficient of b^r, i.e. $\dfrac{n(n-1)(n-2) \ldots (n-r+1)}{r!}$, has r factors in the numerator,

3 the term containing b^2 is the third term, the term in b^3 is the fourth term, and so on,

4 the coefficients are symmetrical, i.e. the coefficient of $a^{n-r}b^r$ is equal to the coefficient of $a^r b^{n-r}$ (you can see this in Pascal's Triangle).

To expand $(1+x)^{10}$ in ascending (i.e. going up) powers of x as far as the term in x^3 we replace n with 10, a with 1 and b with x to give

$$(1+x)^{10} = 1(1)^{10} + 10(1)^9 x + \frac{10 \times 9}{2 \times 1}(1)^8 x^2 + \frac{10 \times 9 \times 8}{3 \times 2 \times 1}(1)^7 x^3 + \ldots + x^{10}$$

$$= 1 + 10x + 45x^2 + 120x^3 + \ldots + x^{10}$$

To expand $(1+x)^{10}$ in descending (i.e. going down) powers of x as far as the term in x^3, first write $(1+x)^{10}$ as $(x+1)^{10}$. Then replace n with 10, a with x and b with 1 to give

$$(1+x)^{10} = (x+1)^{10} = 1(x)^{10} + 10(x)^9(1) + \frac{10 \times 9}{2 \times 1}(x)^8(1)^2 + \frac{10 \times 9 \times 8}{3 \times 2 \times 1}(x)^7(1)^3 + \ldots$$

$$= x^{10} + 10x^9 + 45x^8 + 120x^7 + \ldots + 1$$

The worked examples show how other expressions can be expanded.

Examples 11d

1 Write down the first three terms in the expansion in ascending powers of x of

(a) $\left(1 - \dfrac{x}{2}\right)^{10}$

(b) $(3 - 2x)^8$

(a) Use $(a+b)^n$ and replace a with 1, b with $-\dfrac{x}{2}$ and n with 10.

$$\left(1 - \frac{x}{2}\right)^{10} = 1^{10} + (10)(1)^9\left(-\frac{x}{2}\right) + \frac{10 \times 9}{2 \times 1}(1)^8\left(-\frac{x}{2}\right)^2 + \ldots$$

$$= 1 - 5x + \frac{45}{4}x^2 + \ldots$$

(b) Use $(a+b)^n$ and replace a with 3, b with $-2x$ and n with 8.

$$(3 - 2x)^8 = 3^8 + 8(3)^7(-2x) + \frac{8 \times 7}{2}(3)^6(-2x)^2 + \ldots$$

$$= 3^8 - 16(3^7)x + 112(3^6)x^2 + \ldots$$

2 Find the fourth term in the expansion of $(a - 2b)^{20}$ as a series in ascending powers of b.

The fourth term in the expansion of $(a + b)^n$ is $\binom{n}{3} a^{n-3} b^3$ so replace b with $-2b$ and n with 20.

$$\text{The fourth term is } \binom{20}{3} a^{17}(-2b)^3 = \frac{20 \times 19 \times 18}{3 \times 2 \times 1} a^{17}(-8b^3) = -9120 a^{17} b^3$$

3 Write down the first three terms in the binomial expansion of

$$(1 - 2x)\left(1 + \tfrac{1}{2}x\right)^{10}$$

$$\left(1 + \tfrac{1}{2}x\right)^{10} = 1 + (10)\left(\tfrac{1}{2}x\right) + \frac{(10)(9)}{2!}\left(\tfrac{1}{2}x\right)^2 + \ldots$$

The third term in the binomial expansion is the term containing x^2, so start by expanding $\left(1 + \tfrac{1}{2}x\right)^{10}$ as far as the term in x^2.

$$= 1 + 5x + \tfrac{45}{4}x^2 + \ldots$$

$$\therefore \quad (1 - 2x)\left(1 + \tfrac{1}{2}x\right)^{10} = (1 - 2x)\left(1 + 5x + \tfrac{45}{4}x^2 + \ldots\right)$$

$$= 1 + 5x + \tfrac{45}{4}x^2 + \ldots -2x - 10x^2 + \ldots$$

We do not write down the product of $-2x$ and $\tfrac{45}{4}x^2$, as terms in x^3 are not required.

$$= 1 + 3x + \tfrac{5}{4}x^2 + \ldots$$

Exercise 11d

1 Write down the first four terms in ascending power of x of the binomial expansion of

(a) $(1 + 3x)^{12}$

(b) $(1 - 2x)^9$

(c) $(2 + x)^{10}$

(d) $\left(1 - \dfrac{x}{3}\right)^{20}$

(e) $\left(2 - \dfrac{3}{2x}\right)^7$

(f) $\left(\dfrac{3}{2} + 2x\right)^9$

2 Write down the term indicated in the binomial expansion of each of the following functions.

(a) $(1 - 4x)^7$, third term

(b) $\left(1 - \dfrac{x}{2}\right)^{20}$, second term

(c) $(2 - x)^{15}$, 12th term

(d) $(p - 2q)^{10}$, fifth term

(e) $(3a + 2b)^8$, second term

(f) $(1 - 2x)^{12}$, the term in x^4

(g) $\left(2 + \dfrac{x}{2}\right)^9$, the term in x^5

(h) $(a + b)^8$, the term in a^3

3 Write down the binomial expansion of each function as a series of ascending powers of x as far as, and including, the term in x^2

(a) $(1 + x)(1 - x)^9$ (b) $(1 - x)(1 + 2x)^{10}$

(c) $(2 + x)\left(1 - \dfrac{x}{2}\right)^{20}$ (d) $(1 + x)^2(1 - 5x)^{14}$

4 Find the first three terms and the last term in the expansion of $(1 + 2x)^9$

5 Find the coefficient of x^2 in the expansion of $\left(x + \dfrac{2}{x}\right)^4$

6 The coefficient of x^2 in the expansion of $(1 + 2x^2)(a + x)^{10}$ is 47. Find the value of a given that it is positive.

7 The coefficients of b^2 and b^3 in the expansion of $(a + b)^6$ are equal. Find the value of a.

8 Find the first three terms in the expansion of $(1 + b)^5$. Use the substitution $x = 1 + b$, to find the coefficient of b^2 in the expansion of $(1 + b - 2b^2)^5$

9 There is no term in x in the expansion of $(1 - x)(1 - cx)^{10}$. Find the value of c.

10 Find the coefficient of x^4 in the expansion of $(1 + x^2)(1 + 2x^2)^{10}$

12 Integration

After studying this chapter you should be able to

- understand integration as the reverse process of differentiation, and integrate $(ax + b)^n$ (for any rational n except -1), together with constant multiples, sums and differences
- solve problems involving the evaluation of a constant of integration
- evaluate definite integrals (including simple cases of 'improper' integrals)
- use definite integration to find: the area of a region bounded by a curve and lines parallel to the axes, or between two curves; a volume of revolution about one of the axes.

DIFFERENTIATION REVERSED

When x^2 is differentiated with respect to x the derivative is $2x$.

So if the derivative of an unknown function is $2x$ then the unknown function could be x^2.

This process of finding a function from its derivative, which reverses the operation of differentiating, is called *integration*.

The constant of integration

As seen above, $2x$ is the derivative of x^2, but it is also the derivative of $x^2 + 3$, $x^2 - 9$ and, in fact, the derivative of $x^2 + K$, where K is any constant.

Therefore the result of integrating $2x$, which is called *the integral of* $2x$, is not a unique function but is of the form

$x^2 + K$ where K is any constant

K is called *the constant of integration*.

This is written $\displaystyle\int 2x \, dx = x^2 + K$

where $\displaystyle\int \ldots \, dx$ means 'the integral of … with respect to x'.

Integrating *any* function reverses the process of differentiating, so for any function $f(x)$ we have

$$\int \frac{d}{dx} f(x) \, dx = f(x) + K$$

For example, because differentiating x^3 with respect to x gives $3x^2$ we have $\displaystyle\int 3x^2 \, dx = x^3 + K$

and it follows that $\displaystyle\int x^2 \, dx = \frac{1}{3}x^3 + K$

It is not necessary to write $\frac{1}{3}K$ in the second form, as K represents *any* constant in either expression.

In general, the derivative of x^{n+1} is $(n + 1)x^n$ so $\displaystyle\int x^n \, dx = \frac{1}{n + 1} x^{n+1} + K$

so **to integrate a power of x, *increase* that power by 1 and *divide* by the new power.**

This rule can be used to integrate any power of x *except* -1.

Integrating a sum or difference of functions

We know that a function can be differentiated term by term. Therefore, as integration reverses differentiation, integration also can be done term by term.

Examples 12a

1 Find the integral of $1 + x^7 + \dfrac{1}{x^2} - \sqrt{x}$

$$\int\left(1 + x^7 + \frac{1}{x^2} - \sqrt{x}\right)dx = \int\left(1 + x^7 + x^{-2} - x^{\frac{1}{2}}\right)dx$$

$$= \int 1\,dx + \int x^7\,dx + \int x^{-2}\,dx - \int x^{\frac{1}{2}}\,dx$$

$$= x + \frac{1}{8}x^8 + \frac{1}{-1}x^{-1} - \frac{1}{\frac{3}{2}}x^{\frac{3}{2}} + K$$

$$= x + \frac{1}{8}x^8 - \frac{1}{x} - \frac{2}{3}x^{\frac{3}{2}} + K$$

2 Find $\displaystyle\int \frac{x^2 - x}{\sqrt{x}}\,dx$

$$\int \frac{x^2 - x}{\sqrt{x}}\,dx = \int\left(\frac{x^2}{x^{\frac{1}{2}}} - \frac{x}{x^{\frac{1}{2}}}\right)dx = \int x^{\frac{3}{2}}\,dx - \int x^{\frac{1}{2}}\,dx$$

$$= \frac{2}{5}x^{\frac{5}{2}} - \frac{2}{3}x^{\frac{3}{2}} + K$$

Integrating $(ax + b)^n$

Look at the function $f(x) = (2x + 3)^4$

To differentiate $f(x)$ we make a substitution

i.e. $\qquad u = 2x + 3 \qquad \Rightarrow \qquad f(x) = u^4$

giving $\qquad \dfrac{d}{dx}(2x + 3)^4 = (4)(2)(2x + 3)^3$

Hence $\qquad \displaystyle\int (4)(2)(2x + 3)^3\,dx = (2x + 3)^4 + K$

$\qquad\qquad \displaystyle\int (2x + 3)^3\,dx = \frac{1}{(2)(4)}(2x + 3)^4 + K \qquad\qquad$ dividing by $(2)(4)$

Therefore $f(x) = (ax + b)^{n+1}$ gives the general result

$$\int (ax + b)^n\,dx = \frac{1}{(a)(n + 1)}(ax + b)^{n+1} + K$$

Exercise 12a

Integrate with respect to x

1 x^5

2 $\dfrac{1}{x^5}$

15 $(2 - 3x)(1 + 5x)$

16 $\dfrac{1 + x}{\sqrt{x}}$

3 $\sqrt[4]{x}$

4 x^{-3}

17 $\dfrac{1 - 2x}{x^3}$

18 $\dfrac{1 + x + x^3}{\sqrt{x}}$

5 $\dfrac{1}{x^{\frac{5}{2}}}$

6 $x^{-\frac{1}{2}}$

19 $(1 - x)^2$

20 $x(1 + x)(1 - x)$

7 x^1

8 $\dfrac{1}{\sqrt[3]{x}}$

21 $\dfrac{1 - \sqrt{x}}{x^2}$

22 $(x + 3)^{-2}$

9 $1 + x^2$

10 $x + x^2$

23 $(1 + x)^{\frac{1}{2}}$

24 $(1 + 3x)^5$

11 $2x - \sqrt{x}$

12 $1 + \dfrac{1}{x^2}$

25 $(2 - 5x)^4$

26 $(3x + 5)^3$

13 $\dfrac{x^2 + 1}{x^2}$

14 $x(1 + x)$

27 $(2x - 9)^{-\frac{1}{2}}$

28 $\sqrt{(4x + 1)}$

Finding the constant of integration

When $\dfrac{dy}{dx} = 2x$, we know that $y = x^2 + K$, which is not the equation of a particular curve.

To find the equation of the curve we need to know the coordinates of a point on the curve.

Example 12b

The equation of a curve is such that $\dfrac{dy}{dx} = 3 + 2x$ and $(1, 5)$ is a point on the curve.

Find the equation of the curve.

$$\frac{dy}{dx} = 3 + 2x \quad \Rightarrow \quad y = \int (3 + 2x)\, dx$$

$\therefore \qquad y = 3x + x^2 + K$

$(1, 5)$ is on the curve so $5 = 3 + 1 + K \quad \Rightarrow \quad K = 1$

$\therefore \qquad y = 3x + x^2 + 1$

Exercise 12b

1 The equation of a curve is such that
$$\frac{dy}{dx} = 3x - 4x^3$$
and the curve passes through $(1, 2)$.
Find the equation of the curve.

2 The equation of a curve is such that
$$\frac{dy}{dx} = 2 + 3\sqrt{x}$$
and the curve passes through $(4, -1)$.
Find the equation of the curve.

3 The equation of a curve is such that
$$\frac{dy}{dx} = 1 + \frac{3}{x^2}$$

and the curve passes through $(3, -1)$.
Find the equation of the curve.

4 The equation of a curve is such that
$$\frac{dy}{dx} = 2x(1 + 3x)$$
and the curve passes through $(1, 8)$.
Find the equation of the curve.

5 The equation of a curve is such that
$$\frac{dy}{dx} = (2 - 3x)^3$$
and the point $(2, 10)$ lies on the curve.
Find the equation of the curve.

USING INTEGRATION TO FIND AN AREA

The area shown in the diagram is bounded by the curve $y = f(x)$, the x-axis and the lines $x = a$ and $x = b$

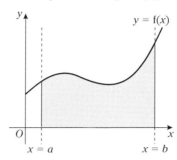

This area can be estimated by dividing the area into thin vertical strips and treating each strip, or *element*, as being approximately rectangular.

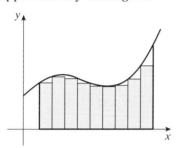 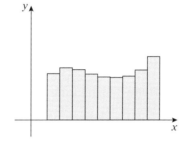

The sum of the areas of the rectangular strips then gives an approximate value for the area. The thinner the strips are, the better the approximation is.

Every strip has one end on the x-axis, one end on the curve and two vertical sides, so all the strips have the same type of boundaries. We call these strips elements.

Look at an element bounded on the left by the vertical line through any point $P(x, y)$ on the curve, then

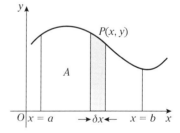

the width of the element represents a small increase in the value of x and so can be called δx.

Also, if A represents the part of the area up to the vertical line through P, then

the area of the element represents a small increase in the value of A and so can be called δA.

The element is approximately a rectangle of height y and width δx.

Therefore, for any element $\delta A \approx y \, \delta x$

The equation $\delta A \approx y \, \delta x$ can be written as $\dfrac{\delta A}{\delta x} \approx y$

This becomes more accurate as δx gets smaller, giving $\displaystyle\lim_{\delta x \to 0} \dfrac{\delta A}{\delta x} = y$

But $\displaystyle\lim_{\delta x \to 0} \dfrac{\delta A}{\delta x}$ is $\dfrac{\mathrm{d}A}{\mathrm{d}x}$ so $\dfrac{\mathrm{d}A}{\mathrm{d}x} = y$

Hence $A = \displaystyle\int y \, \mathrm{d}x$

The boundary values of x defining the total area are $x = a$ and $x = b$ and we show this by writing

$$\textbf{total area} = \int_a^b \boldsymbol{y} \, \mathbf{d}\boldsymbol{x}$$

DEFINITE INTEGRATION

The area bounded by the x-axis, the lines $x = a$ and $x = b$ and the curve $y = 3x^2$ is given by $A = \int 3x^2 \, dx$, i.e. $A = x^3 + K$

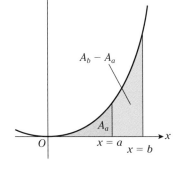

From this area function we can find the value of A corresponding to a particular value of x.

Hence   using $x = a$ gives $A_a = a^3 + K$

and   using $x = b$ gives $A_b = b^3 + K$

Then the area between $x = a$ and $x = b$ is given by $A_b - A_a$ where
$A_b - A_a = (b^3 + K) - (a^3 + K) = b^3 - a^3$

$A_b - A_a$ is called the definite integral from a to b of $3x^2$ and is denoted by

$$\int_a^b 3x^2 \, dx, \quad \text{i.e.} \quad \int_a^b 3x^2 \, dx = (x^3)_{x=b} - (x^3)_{x=a}$$

The right-hand side of this equation is usually written in the form $\left[x^3\right]_a^b$ where a and b are called the *boundary values* or *limits of integration*; b is the *upper limit* and a is the *lower limit*.

$\int_a^b y \, dx$ is called the definite integral from a to b of y with respect to x.

Whenever a definite integral is calculated, the constant of integration disappears.

A definite integral can be found in this way only if the function to be integrated is defined for every value of x from a to b, for example

$\int_{-1}^{1} \frac{1}{x^2} \, dx$ cannot be found directly as $\frac{1}{x^2}$ is undefined when $x = 0$

Example 12c

Evaluate $\displaystyle\int_1^4 \frac{1}{x^2} \, dx$

$$\int_1^4 \frac{1}{x^2} \, dx \quad \equiv \int_1^4 x^{-2} \, dx$$

$$= \left[-x^{-1} \right]_1^4 = \{ -4^{-1} \} - \{ -1^{-1} \} = -\tfrac{1}{4} + 1 = \tfrac{3}{4}$$

Exercise 12c

Evaluate each of the following definite integrals.

1 $\displaystyle\int_0^2 x^3 \, dx$ **2** $\displaystyle\int_1^2 \sqrt{x^5} \, dx$ **7** $\displaystyle\int_{-1}^0 (1 - x)^2 \, dx$ **8** $\displaystyle\int_1^2 \frac{3x + 1}{\sqrt{x}} \, dx$

3 $\displaystyle\int_2^4 (x^2 + 4) \, dx$ **4** $\displaystyle\int_4^9 \sqrt{x} \, dx$ **9** $\displaystyle\int_{-1}^0 (2 + 3x)^2 \, dx$ **10** $\displaystyle\int_{\frac{1}{2}}^1 (3 - 2x)^3 \, dx$

5 $\displaystyle\int_0^3 (x^2 + 2x - 1) \, dx$ **6** $\displaystyle\int_0^2 (x^3 - 3x) \, dx$

FINDING AREA BY DEFINITE INTEGRATION

The area bounded by a curve $y = f(x)$, the lines $x = a$, $x = b$, and the x-axis, can be found from the definite integral

$$\int_a^b f(x)\ dx$$

Example 12d

Find the area of the shaded region shown in the diagram.

The required area starts at the y-axis, i.e. at $x = 0$ and ends where the curve crosses the x-axis, i.e. where $x = 1$

$$\text{Area} = \int_0^1 (1 - x^2)\ dx = \left[x - \frac{x^3}{3} \right]_0^1 = \left(1 - \frac{1}{3} \right) - (0 - 0) = \frac{2}{3}$$

The area is $\frac{2}{3}$ of a square unit.

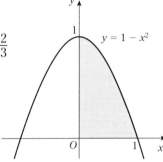

Exercise 12d

In each question find the area with the given boundaries.

1 The x-axis, the curve $y = x^2 + 3$ and the lines $x = 1$, $x = 2$

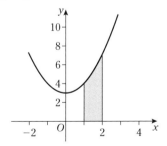

2 The curve $y = \sqrt{x}$, the x-axis and the lines $x = 4$, $x = 9$

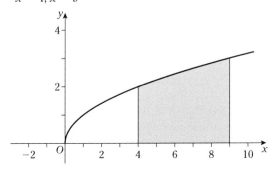

3 The x-axis, the lines $x = -1$, $x = 1$, and the curve $x^2 + 1$.

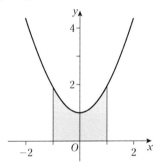

4 The curve $y = x^2 + x$, the x-axis and the line $x = 3$

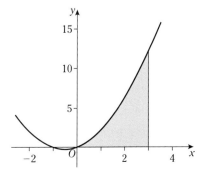

5 The positive x- and y-axes and the curve $y = 4 - x^2$

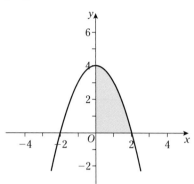

6 The lines $x = 2$, $x = 4$, the x-axis and the curve $y = x^3$

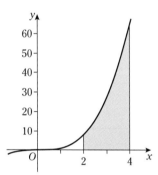

7 The curve $y = 4 - x^2$, the positive y-axis and the negative x-axis.

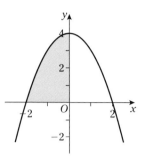

8 The x-axis, the lines $x = 1$ and $x = 2$, and the curve $y = \frac{1}{2}x^3 + 2x$

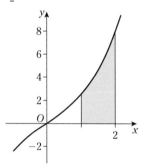

9 The x-axis and the lines $x = 1$, $x = 5$ and $y = 2x$. Check the result by sketching the required area and finding it by calculating the area of a trapezium.

The meaning of a negative result

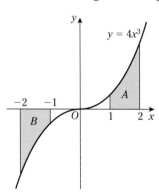

Look at the area bounded by $y = 4x^3$, the x-axis and the lines

(a) $x = 1$ and $x = 2$ (b) $x = -2$ and $x = -1$

This curve is symmetrical about the origin so the two shaded areas are equal.

(a) The area of A is

$$\int_1^2 y\, dx = \int_1^2 4x^3\, dx$$

$$= \left[x^4 \right]_1^2 = 16 - 1 = 15$$

(b) The area of B is

$$\int_{-2}^{-1} y\, dx = \int_{-2}^{-1} 4x^3\, dx$$

$$= \left[x^4 \right]_{-2}^{-1}$$

$$= 1 - 16$$

$$= -15$$

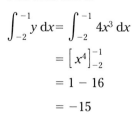

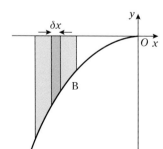

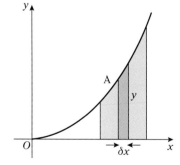

This integral has a negative value because, from -1 to -2, the value of y that gives the length of the strip is negative. An area cannot be negative so the minus sign just means that the area is below the x-axis. The actual area is 15 square units.

Example 12e

Find the area enclosed between the curve $y = x(x - 1)(x - 2)$ and the x-axis.

The curve crosses the x-axis at $x = 0$, $x = 1$ and $x = 2$. The area enclosed between the curve and the x-axis is the sum of the areas A and B.

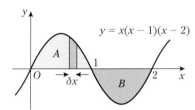

The area of A is $\displaystyle\int_0^1 y \, dx = \int_0^1 (x^3 - 3x^2 + 2x) \, dx$

$$= \left[\frac{x^4}{4} - x^3 + x^2 \right]_0^1 = \frac{1}{4}$$

The area of B is $\displaystyle\int_1^2 y \, dx = \int_1^2 (x^3 - 3x^2 + 2x) \, dx$

$$= \left[\frac{x^4}{4} - x^3 + x^2 \right]_1^2 = (4 - 8 + 4) - \left(\frac{1}{4} - 1 + 1 \right) = -\frac{1}{4}$$

The negative sign means that the area of B is below the x-axis. The actual area of B is $\frac{1}{4}$ of a square unit.

The total area is $\left(\frac{1}{4} + \frac{1}{4} \right)$ square unit $= \frac{1}{2}$ square unit

Exercise 12e

1 Find the area below the x-axis and above the curve $y = x^2 - 1$

2 Find the area bounded by the curve $y = 1 - x^3$, the x-axis and the lines $x = 2$ and $x = 3$

3 Find the area between the x and y axes and the curve $y = (1 - x)^2$

4 Sketch the curve $y = x(x^2 - 1)$, showing where it crosses the x-axis.

Find

   (a) the area enclosed above the x-axis and below the curve

   (b) the area below the x-axis and above the curve

   (c) the total area between the curve and the x-axis.

5 Repeat question 4 for the curve $y = x(4 - x^2)$

6 Find

   (a) $\displaystyle\int_0^2 (x - 2) \, dx$ (b) $\displaystyle\int_2^4 (x - 2) \, dx$

   (c) $\displaystyle\int_4^0 (x - 2) \, dx$

Interpret your results with a sketch.

Finding the area between a curve and the y-axis

The area between a curve and the y-axis can be found by subtracting the area between a curve and the x-axis from the area of a rectangle. The next worked example shows how to do this.

Examples 12f

1 Find the area enclosed by the curve $y = x^3 + 1$, the y-axis and the line $y = 9$

$y = x^3 + 1$ and $y = 9$ intersect where $9 = x^3 + 1$, i.e. where $x = 2$

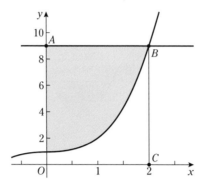

Area between the curve, the x-axis, $x = 0$ and $x = 2$ is

$$\int_0^2 (x^3 + 1)\, dx$$

and the area of rectangle $OABC$ is 18.

Therefore the shaded area is

$$18 - \int_0^2 (x^3 + 1)\, dx \qquad = 18 - \left[\frac{x^4}{4} + x \right]_0^2$$

$$= 18 - 6 = 12$$

Finding compound areas

Compound areas can also be found by subtracting one area from another.

Examples 12f cont.

2 Find the area between the curve $y = x^2$ and the line $y = 3x$ as shown in the diagram.

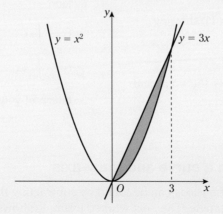

The line and curve meet where $x^2 = 3x$, i.e. where $x = 0$ and $x = 3$

Area between the curve, the x-axis, $x = 0$ and $x = 3$ is $\int_0^3 x^2 \, \mathrm{d}x$

and the area between the line, the x-axis, $x = 0$ and $y = 3$ is $\int_0^3 3x \, \mathrm{d}x$

$\therefore$ the shaded area is $\int_0^3 3x \, \mathrm{d}x - \int_0^3 x^2 \, \mathrm{d}x = \int_0^3 (3x - x^2) \, \mathrm{d}x$

$$= \left[\tfrac{3}{2}x^2 - \tfrac{1}{3}x^3 \right]_0^3 = \tfrac{27}{2} - 9 = 4\tfrac{1}{2}$$

3 Find the area between the curve $y = x^3$ and $y = 4x - 3x^2$ shown in the diagram.

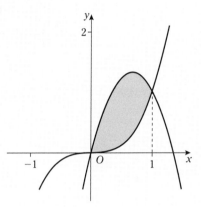

The curves meet where $\qquad x^3 = 4x - 3x^2$

$\Rightarrow \qquad\qquad x^3 + 3x^2 - 4x = 0$

$\Rightarrow \qquad\qquad x(x^2 + 3x - 4) = 0$

$\Rightarrow \qquad x = 0 \quad \text{or} \quad x = 1 \quad \text{or} \quad x = -4$

The diagram shows the area between the curves from $x = 0$ to $x = 1$

The area between $y = x^3$, the x-axis and the line $x = 1$ is $\int_0^1 x^3 \, \mathrm{d}x$

and the area between $y = 4x - 3x^2$, the x-axis and the line $x = 1$ is $\int_0^1 (4x - 3x^2) \, \mathrm{d}x$.

The required area is

$$\int_0^1 (4x - 3x^2) \, \mathrm{d}x - \int_0^1 x^3 \, \mathrm{d}x = \int_0^1 (4x - 3x^2 - x^3) \, \mathrm{d}x = \left[2x^2 - x^3 - \tfrac{1}{4}x^4 \right]_0^1 = \tfrac{3}{4}$$

Exercise 12f

1 Find the shaded area in the diagram that is enclosed by the curve $y = x(2 - x)$, the y-axis and the line $y = 1$

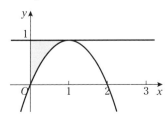

2 Calculate the area bounded by the curve $y = \sqrt{x}$, the y-axis and the line $y = 3$

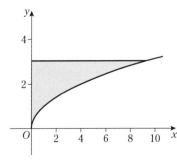

3 Find the area bounded by

(a) the x-axis, the line $x = 2$ and the curve $y = x^2$

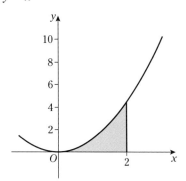

(b) the y-axis, the line $y = 4$ and the curve $y = x^2$

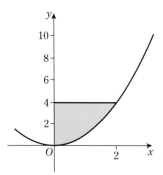

4 A region in the xy-plane is bounded by the x-axis, the y-axis, the lines $y = 1$ and $x = 2$ and the curve $y = \dfrac{1}{x^2}$

Find its area.

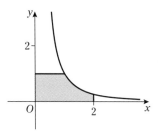

5 Find the area between the curves $y = 1 - x^2$ and $y = 1 - x$

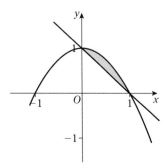

6 Calculate the area in the first quadrant between the curve $y^2 = x$ and the lines $x = 9, y = 0$

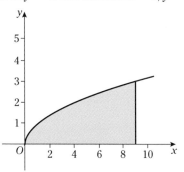

7 Find the area between the x-axis, $x = -1$ and the curve $y = (1 - x)^{\frac{1}{2}}$

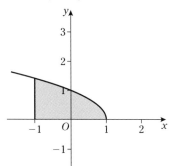

8 Evaluate the area between the line $y = x - 1$ and the curve

(a) $y = x(1 - x)$

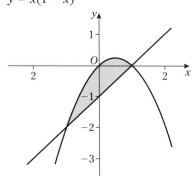

(b) $y = (2x + 1)(x - 1)$

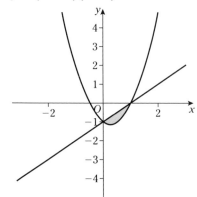

9　Calculate the area of the region of the xy-plane between the curves $y = (x + 1)(x - 2)$ and $y = x$

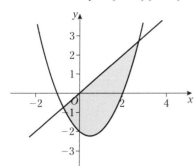

10　Calculate the area between the curves $y = x^2$ and $y = x(2 - x)$

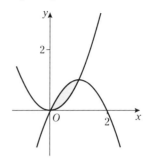

Improper integrals

In this diagram the shaded area is bounded by the curve $y = f(x)$, the x-axis and the two lines $x = a$ and $x = b$. Both vertical lines intersect the curve, so the area is clearly defined and can be evaluated using $\displaystyle\int_a^b f(x)\,dx$.

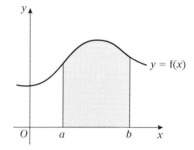

But the area may not always be so clear.

For example, the area bounded by the curve $y = \dfrac{1}{x^2}$, the x-axis and the line $x = 1$. In this case $a = 1$ and, as the x-axis is an asymptote to the curve, b is infinitely large and therefore indeterminate.

We cannot be sure what happens to the area as $x \to \infty$, so we will first take a large, *finite* value for b, $b = X$ say.

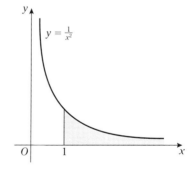

Finding this area gives

$$\int_1^X \frac{1}{x^2}\,dx = \left[\frac{-1}{x}\right]_1^X \quad \Rightarrow \quad -\frac{1}{X} + 1$$

Now as X approaches infinity, $\dfrac{1}{X}$ approaches 0.

Therefore the value of the area converges to 1, and we say that

$$\int_1^\infty \frac{1}{x^2}\,dx \quad \text{is convergent and equal to 1.}$$

Now consider the area bounded by the same curve $y = \dfrac{1}{x^2}$, the y-axis and the line $x = 1$

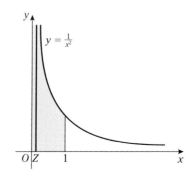

This time the y-axis is an asymptote, and we are not sure what happens to the area as $x \to 0$, so we take a very small value for a, $a = Z$ say.

Integrating as before gives

$$\int_Z^1 \frac{1}{x^2}\,dx = \left[\frac{-1}{x}\right]_Z^1 \quad \Rightarrow \quad -1 + \frac{1}{Z}$$

As $Z \to 0$, $\dfrac{1}{Z} \to \infty$, so the value of the area becomes infinitely large, so $\displaystyle\int_0^1 \frac{1}{x^2}\,dx$ is indeterminate.

Each of these cases is an example of an *improper integral* and they show that some, but not all, improper integrals can be found. Any definite integral that has one of its limits at a value of x where there is an asymptote, is an improper integral, whether or not the limit is zero or infinity.

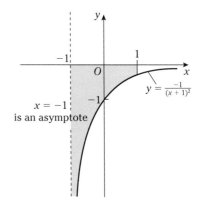

$x = -1$ is an asymptote

An example of this type is $\int_{-1}^{1} \frac{-1}{(x+1)^2} \, dx$.

Exercise 12g

Find the value, if it exists, of each of the following integrals.

1 $\int_{1}^{\infty} \frac{1}{x^3} \, dx$

2 $\int_{0}^{1} \frac{1}{x^3} \, dx$

3 $\int_{0}^{2} \frac{1}{\sqrt{x}} \, dx$

4 $\int_{2}^{\infty} \frac{1}{\sqrt{x}} \, dx$

5 $\int_{1}^{4} \frac{1}{(1 - x^2)} \, dx$

VOLUME OF REVOLUTION

When an area is rotated about a straight line, the three-dimensional object formed is called a *solid of revolution*, and its volume is a *volume of revolution*.

The line about which rotation takes place is always an axis of symmetry for the solid of revolution. Also, any cross-section of the solid that is perpendicular to the axis of rotation is circular.

Look at the solid of revolution formed when the shaded area is rotated about the x-axis.

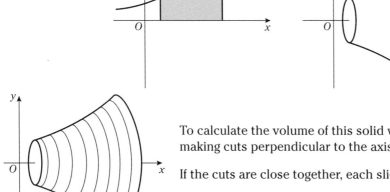

To calculate the volume of this solid we can divide it into 'slices' by making cuts perpendicular to the axis of rotation.

If the cuts are close together, each slice is approximately a cylinder.

Think about an element formed by one cut through the point $P(x, y)$ and the other cut distant δx from the first.

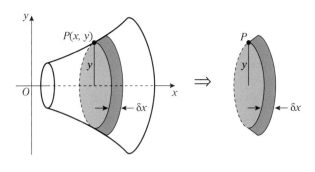

The volume, δV, of this element is approximately that of a cylinder of radius y and 'height' δx,

i.e. $\delta V \approx \pi y^2\, \delta x$ $\Rightarrow$ $\dfrac{\delta V}{\delta x} \approx \pi y^2$

giving $\dfrac{\mathrm{d}V}{\mathrm{d}x} = \pi y^2$

Therefore $V = \displaystyle\int \pi y^2\, \mathrm{d}x$

When the equation of the rotated curve is $y = f(x)$, this integral can be evaluated. For example when $y = x^2$, $V = \displaystyle\int \pi (x^2)^2\, \mathrm{d}x$

When the area is rotated about the y-axis, we can use a similar method based on slices perpendicular to the y-axis, giving $V = \displaystyle\int \pi x^2\, \mathrm{d}y$

Examples 12h

1 Find the volume generated when the area between $y = x^2$, the line $x = 2$ and the x-axis is rotated about the x-axis.

We are integrating with respect to x so the limits of the integral are the limits of x, i.e. $x = 0$ and $x = 2$

$$V = \int_0^2 \pi y^2\, \mathrm{d}x = \int_0^2 \pi x^4\, \mathrm{d}x$$

$$= \left[\frac{1}{5}\pi x^5 \right]_0^2 = \frac{32\pi}{5} = 20.1 \ (3\ \text{s.f.})$$

Therefore the volume generated is 20.1 cubic units.

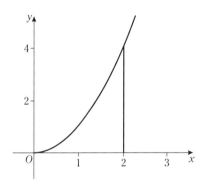

2 The area defined by the inequalities $y \geqslant x^2 + 1$, $x \geqslant 0$, $y \leqslant 2$ is rotated completely about the y-axis. Find the volume of the solid generated.

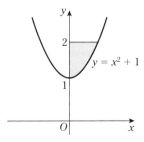
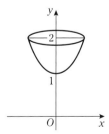

Rotating the shaded area about the y-axis gives the solid shown. This time we use horizontal cuts to form approximate cylinders with radius x and thickness δy.

$\delta V = \pi x^2 \delta y \Rightarrow \dfrac{\delta V}{\delta y} = \pi x^2$

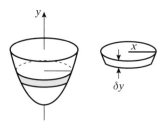

$\dfrac{\mathrm{d}V}{\mathrm{d}y} = \pi x^2 \qquad \text{so} \qquad V = \displaystyle\int_1^2 \pi x^2 \,\mathrm{d}y$

We cannot integrate x^2 with respect to y so we need to express x^2 in terms of y.

Using the equation of the curve, $y = x^2 + 1$ gives $x^2 = y - 1$

$$V = \pi \int_1^2 (y - 1)\,\mathrm{d}y = \pi\left[\tfrac{1}{2}y^2 - y\right]_1^2$$
$$= \pi\left[(2 - 2) - \left(\tfrac{1}{2} - 1\right)\right] = \tfrac{1}{2}\pi$$

The volume of the solid is $\tfrac{1}{2}\pi$ cubic units.

3 Find the volume generated when the area between the curve $y^2 = x$ and the line $y = x$ is rotated completely about the x-axis.

The defined area is shown in the diagram.

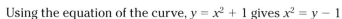

When this area rotates about Ox, the solid generated is bowl-shaped on the outside, with a conical hole inside. The cross-section this time is not a simple circle but is an annulus, i.e. the area between two concentric circles.

For a typical element the area of cross-section is $\pi y_1^2 - \pi y_2^2$

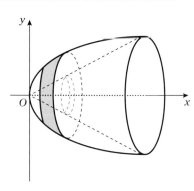

Therefore the volume of an element is given by
$\delta V \approx \pi\left\{y_1^2 - y_2^2\right\}\delta x$

$\Rightarrow \qquad \dfrac{\mathrm{d}V}{\mathrm{d}x} = \pi(y_1^2 - y_2^2)$

$\therefore \qquad V = \pi\displaystyle\int_0^1 (y_1^2 - y_2^2)\,\mathrm{d}x$

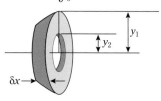

$y_1 = \sqrt{x}$ and $y_2 = x$

$\therefore \qquad V = \pi\displaystyle\int_0^1 (x - x^2)\,\mathrm{d}x = \pi\left[\tfrac{1}{2}x^2 - \tfrac{1}{3}x^3\right]_0^1 = \tfrac{1}{6}\pi$

The volume generated is $\tfrac{1}{6}\pi$ cubic units.

The volume in Example 3 can also be found by calculating separately

(1) the volume given when the curve $y^2 = x$ rotates about the x-axis;

(2) the volume of a cone with base radius 1 and height 1 and subtracting it from (1).

The method in which an annulus element is used, however, applies whatever the shape of the hollow interior.

Exercise 12h

In each of the following questions, find the volume generated when the area shown by the following diagrams is rotated completely about the x-axis.

1

$y = x(4 - x)$

2

$y = \frac{1}{x}$

3

$y = x^2$

4

$y = \pm\sqrt{x}$

In each of the following questions, the area bounded by the curve and line(s) given is rotated about the y-axis to form a solid. Find the volume generated.

5 $y = x^2$, $y = 4$

6 $y = 4 - x^2$, $y = 0$

7 $y = x^3$, $y = 1$, $y = 2$, for $x \geqslant 0$

8 Find the volume generated when the area enclosed between $y = \sqrt{x}$ and $x = 1$ is rotated about the x-axis.

9 (a) The area enclosed between the line $y = x^2$ and the line $y = 2x$ is rotated about the x-axis.

Find the volume generated.

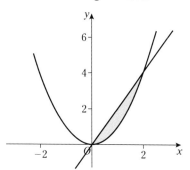

(b) Find the volume generated when the area given in (a) is rotated about the y-axis.

Mixed exercise 12

Integrate with respect to x

1 $x^2 - \dfrac{1}{x^2}$

2 $\sqrt[3]{x}$

3 $\sqrt{x} + \dfrac{1}{\sqrt{x}}$

4 $\dfrac{x^3 - 1}{x^2}$

5 $\dfrac{x^2 - 1}{\sqrt{x}}$

Evaluate

6 $\displaystyle\int_3^6 (6 - x)^2 \, dx$

7 $\displaystyle\int_1^{32} \left(\sqrt[5]{x} - \dfrac{1}{\sqrt[5]{x}} \right) dx$

Find the areas specified in questions **8** to **10**.

8 Bounded by the x- and y-axes and the curve $y = 1 - x^3$

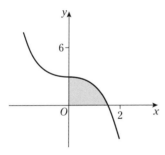

9 Bounded by the curve $y = (x + 4)^{\frac{1}{2}}$ and the x- and y-axes.

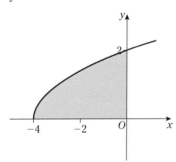

10 The area between the curve $y = (x - 1)(x - 2) + 1$ and the line $2y = x + 1$

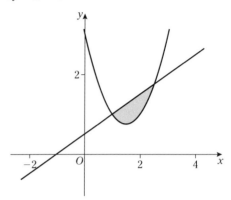

11 Find the volume generated when the area defined in question **8** is rotated about the x-axis.

12 (a) Find the volume generated when the area defined in question **9** is rotated about the x-axis.

   (b) Find the volume generated when the area defined in question **9** is rotated about the y-axis.

13 Evaluate $\displaystyle\int_1^{\infty} \dfrac{1}{(1 + x)^2} \, dx$.

13 Vectors

After studying this chapter you should be able to

- use standard notations for vectors, i.e. $\begin{pmatrix} x \\ y \end{pmatrix}$, $x\mathbf{i} + y\mathbf{j}$, $\begin{pmatrix} x \\ y \\ z \end{pmatrix}$, $x\mathbf{i} + y\mathbf{j} + z\mathbf{k}$, $\overrightarrow{AB}$, $\mathbf{a}$

- carry out addition and subtraction of vectors and multiplication of a vector by a scalar, and interpret these operations in geometrical terms
- use unit vectors, displacement vectors and position vectors
- calculate the magnitude of a vector and the scalar product of two vectors
- use the scalar product to determine the angle between two directions and to solve problems concerning perpendicularity of vectors.

VECTORS

A vector is a quantity which has both magnitude and a specific direction in space.

A scalar quantity is one that is fully defined by magnitude alone. Length, for example, is a scalar quantity, as the length of a piece of string does not depend on its direction when it is measured.

A vector can be represented by a straight line segment where the length of the line represents the magnitude and the direction of the line represents the direction of the vector.

The vector can be denoted by $\overrightarrow{AB}$, where A and B are the end points of the line and the arrow shows the direction, i.e. from A to B. A vector in the opposite direction is denoted by $\overrightarrow{BA}$. The vector can also be denoted by, for example, $\mathbf{a}$.

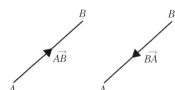

PROPERTIES OF VECTORS

The modulus of a vector

The *modulus* of a vector $\mathbf{a}$ is its magnitude and is written $|\mathbf{a}|$ or a i.e. $|\mathbf{a}|$ is the length of the line representing $\mathbf{a}$.

Equal vectors

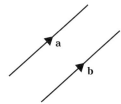

Two vectors with the same magnitude and the same direction are equal.

i.e. $\mathbf{a} = \mathbf{b} \Leftrightarrow \begin{cases} |\mathbf{a}| = |\mathbf{b}| \text{ and} \\ \text{the directions of } \mathbf{a} \text{ and } \mathbf{b} \text{ are the same.} \end{cases}$

It follows that a vector can be represented by *any* line of the right length and direction, regardless of position, i.e. each of the lines in the diagram opposite represents the vector $\mathbf{a}$.

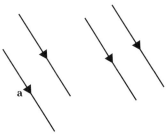

Negative vectors

If two vectors, **a** and **b**, have the same magnitude but opposite directions then

b = −**a**

i.e. −**a** is a vector of magnitude |**a**| and in the direction opposite to that of **a**.

a and **b** are called *equal and opposite* vectors.

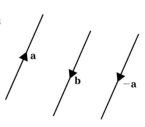

Multiplication of a vector by a scalar

If λ is a positive real number, then λ**a** is a vector in the same direction as **a** and of magnitude λ|**a**|.

It follows that $-\lambda$**a** is a vector in the opposite direction, with magnitude λ|**a**|.

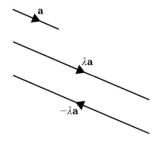

Addition of vectors

If the sides *AB* and *BC* of a triangle *ABC* represent the vectors **p** and **q** then the third side *AC* represents the vector sum, or resultant, of **p** and **q**, which is denoted by **p** + **q**.

Note that **p** and **q** follow each other round the triangle (in this case in the clockwise sense), whereas the resultant, **p** + **q**, goes the opposite way round (anticlockwise in the diagram).

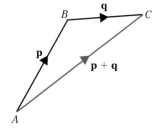

The order in which the addition is performed does not matter, as we can see from a parallelogram *ABCD*.

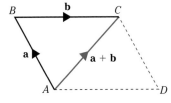
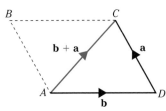

Because the opposite sides of a parallelogram are equal and parallel, $\overrightarrow{AB}$ and $\overrightarrow{DC}$ both represent **a** and $\overrightarrow{BC}$ and $\overrightarrow{AD}$ both represent **b**.

In $\triangle ABC$ $\qquad\qquad \overrightarrow{AC} = \mathbf{a} + \mathbf{b}$ $\qquad$ and in $\triangle ADC$ $\qquad\qquad \overrightarrow{AC} = \mathbf{b} + \mathbf{a}$

Therefore **a** + **b** = **b** + **a**

POSITION VECTORS AND DISPLACEMENT VECTORS

Usually a vector has no specific location in space and is called a *displacement vector*. Some vectors, however, are constrained to a specific position, e.g. the vector $\overrightarrow{OA}$ where *O* is a fixed origin.

$\overrightarrow{OA}$ gives the position vector of *A* relative to *O*.

This displacement is unique and *cannot* be represented by any other line of equal length and direction.

Vectors such as $\overrightarrow{OA}$, representing quantities that have a specific location relative to the origin, are called *position* vectors.

A vector that gives the displacement of one point relative to another point that is not the origin is called a *displacement vector.*

$\overrightarrow{OA}$ gives the displacement of A from O so $\overrightarrow{OA}$ is the position vector of A.

$\overrightarrow{AB}$ is a displacement vector as it gives the displacement of B from A.

In this diagram, O is the origin.

THE LOCATION OF A POINT IN THREE DIMENSIONS

Any point P in a plane can be located by giving its distances from a fixed point O, in each of two perpendicular directions. These distances are the Cartesian coordinates of the point.

To locate a point in three dimensions we start from a fixed point, O. Any other point can be located by giving its distances from O in each of *three* mutually perpendicular directions, i.e. we need *three* coordinates to locate a point in 3-D. So we use the familiar x- and y-axes, together with a third axis Oz. Then any point has coordinates (x, y, z) relative to the origin O.

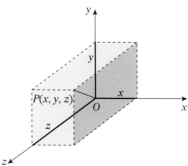

Cartesian unit vectors

A unit vector is a vector whose magnitude is one unit.

Now when

i is a unit vector in the direction of Ox

j is a unit vector in the direction of Oy

k is a unit vector in the direction of Oz

then the position vector, relative to O, of any point P can be given in terms of **i**, **j** and **k**,

e.g. the point distant

3 units from O in the direction of Ox

4 units from O in the direction of Oy

5 units from O in the direction of Oz

has coordinates $(3, 4, 5)$ and $\overrightarrow{OP} = 3\mathbf{i} + 4\mathbf{j} + 5\mathbf{k}$

This can also be written as $\overrightarrow{OP} = \begin{pmatrix} 3 \\ 4 \\ 5 \end{pmatrix}$

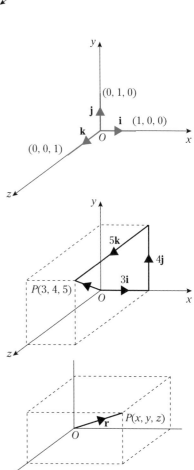

When P is any point (x, y, z) and $\overrightarrow{OP} = \mathbf{r}$

then $\mathbf{r} = x\mathbf{i} + y\mathbf{j} + z\mathbf{k}$

or $\mathbf{r} = \begin{pmatrix} x \\ y \\ z \end{pmatrix}$

and **r** is the position vector of P.

Displacement vectors can be given in the same way.
For example, the vector $3\mathbf{i} + 4\mathbf{j} + 5\mathbf{k}$ *can* represent the position vector of the point $P(3, 4, 5)$ but it can equally well represent *any* vector of length and direction equal to those of $\overrightarrow{OP}$.

Unless a vector is specified as a position vector it is taken to be a displacement vector.

OPERATIONS ON CARTESIAN VECTORS

Addition and subtraction

To add or subtract vectors given in $\mathbf{i}\,\mathbf{j}\,\mathbf{k}$ form, the coefficients of $\mathbf{i}$, $\mathbf{j}$ and $\mathbf{k}$ are added or subtracted separately,

e.g.   if $\mathbf{v}_1 = 3\mathbf{i} + 2\mathbf{j} + 2\mathbf{k}$ and $\mathbf{v}_2 = \mathbf{i} + 2\mathbf{j} - 3\mathbf{k}$

then $\mathbf{v}_1 + \mathbf{v}_2 = (3\mathbf{i} + 2\mathbf{j} + 2\mathbf{k}) + (\mathbf{i} + 2\mathbf{j} - 3\mathbf{k})$

$$= (3 + 1)\mathbf{i} + (2 + 2)\mathbf{j} + (2 - 3)\mathbf{k}$$

$$= 4\mathbf{i} + 4\mathbf{j} - \mathbf{k}$$

and $\mathbf{v}_1 - \mathbf{v}_2 = (3 - 1)\mathbf{i} + (2 - 2)\mathbf{j} + (2 - \{-3\})\mathbf{k}$

$$= 2\mathbf{i} + 5\mathbf{k}$$

Modulus

The modulus of $\mathbf{v}$, where $\mathbf{v} = 12\mathbf{i} - 3\mathbf{j} + 4\mathbf{k}$, is the length of OP where P is the point $(12, -3, 4)$.

Using Pythagoras twice we have

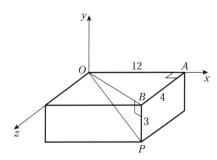

$$OB^2 = OA^2 + AB^2 = 12^2 + 4^2$$

$$OP^2 = OB^2 + BP^2 = (12^2 + 4^2) + (-3)^2$$

$$\therefore \quad OP = \sqrt{12^2 + 4^2 + 3^2} = 13$$

For any vector $\mathbf{v} = a\mathbf{i} + b\mathbf{j} + c\mathbf{k}$

$$|\mathbf{v}| = \sqrt{a^2 + b^2 + c^2}$$

Parallel vectors

Two vectors $\mathbf{v}_1$ and $\mathbf{v}_2$ are parallel if $\mathbf{v}_1 = \lambda\mathbf{v}_2$

e.g. $2\mathbf{i} - 3\mathbf{j} - \mathbf{k}$ is parallel to $4\mathbf{i} - 6\mathbf{j} - 2\mathbf{k}$ $(\lambda = 2)$

and $\mathbf{i} + \mathbf{j} + \mathbf{k}$ is parallel to $-3\mathbf{i} - 3\mathbf{j} - 3\mathbf{k}$ $(\lambda = -3)$

Equal vectors

When two vectors $\mathbf{v}_1 = a_1\mathbf{i} + b_1\mathbf{j} + c_1\mathbf{k}$ and $\mathbf{v}_2 = a_2\mathbf{i} + b_2\mathbf{j} + c_2\mathbf{k}$ are equal then

$$a_1 = a_2 \quad \text{and} \quad b_1 = b_2 \quad \text{and} \quad c_1 = c_2$$

Examples 13a

1 Given the vector $\mathbf{v}$ where $\mathbf{v} = 5\mathbf{i} - 2\mathbf{j} + 4\mathbf{k}$, state whether each of the following vectors is parallel to $\mathbf{v}$, equal to $\mathbf{v}$ or neither.

  (a) $10\mathbf{i} - 4\mathbf{j} + 8\mathbf{k}$

  (b) $-\frac{1}{2}(-10\mathbf{i} + 4\mathbf{j} - 8\mathbf{k})$

  (c) $-5\mathbf{i} + 2\mathbf{j} - 4\mathbf{k}$

  (d) $4\mathbf{i} - 2\mathbf{j} + 5\mathbf{k}$

(a) $10\mathbf{i} - 4\mathbf{j} + 8\mathbf{k} = 2(5\mathbf{i} - 2\mathbf{j} + 4\mathbf{k})$ $(\lambda = 2)$

     $\therefore$ $10\mathbf{i} - 4\mathbf{j} + 8\mathbf{k}$ is parallel to $\mathbf{v}$.

(b) $-\frac{1}{2}(-10\mathbf{i} + 4\mathbf{j} - 8\mathbf{k}) = 5\mathbf{i} - 2\mathbf{j} + 4\mathbf{k}$

     $\therefore$ $-\frac{1}{2}(-10\mathbf{i} + 4\mathbf{j} - 8\mathbf{k})$ is equal to $\mathbf{v}$.

(c) $-5\mathbf{i} + 2\mathbf{j} - 4\mathbf{k} = -(5\mathbf{i} - 2\mathbf{j} + 4\mathbf{k})$ $(\lambda = -1)$

     $\therefore$ $-5\mathbf{i} + 2\mathbf{j} - 4\mathbf{k}$ is parallel to $\mathbf{v}$.

(d) $4\mathbf{i} - 2\mathbf{j} + 5\mathbf{k}$ is not a multiple of $5\mathbf{i} - 2\mathbf{j} + 4\mathbf{k}$.

     $\therefore$ $4\mathbf{i} - 2\mathbf{j} + 5\mathbf{k}$ is not equal or parallel to $\mathbf{v}$.

2 A triangle ABC has its vertices at the points $A(2, -1, 4)$, $B(3, -2, 5)$ and $C(-1, 6, 2)$.

Find, in the form $a\mathbf{i} + b\mathbf{j} + c\mathbf{k}$, the vectors $\overrightarrow{AB}$, $\overrightarrow{BC}$ and $\overrightarrow{CA}$ and hence find the perimeter of the triangle.

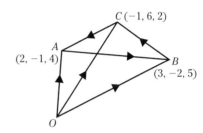

$$\overrightarrow{AB} = \overrightarrow{AO} + \overrightarrow{OB}$$

$\therefore$ $\overrightarrow{AB} = \overrightarrow{OB} - \overrightarrow{OA}$

          $= (3\mathbf{i} - 2\mathbf{j} + 5\mathbf{k}) - (2\mathbf{i} - \mathbf{j} + 4\mathbf{k})$

          $= \mathbf{i} - \mathbf{j} + \mathbf{k}$

     $\overrightarrow{BC} = \overrightarrow{OC} - \overrightarrow{OB}$

          $= (-\mathbf{i} + 6\mathbf{j} + 2\mathbf{k}) - (3\mathbf{i} - 2\mathbf{j} + 5\mathbf{k})$

          $= -4\mathbf{i} + 8\mathbf{j} - 3\mathbf{k}$

     $\overrightarrow{CA} = \overrightarrow{OA} - \overrightarrow{OC}$

          $= (2\mathbf{i} - \mathbf{j} + 4\mathbf{k}) - (-\mathbf{i} + 6\mathbf{j} + 2\mathbf{k})$

          $= 3\mathbf{i} - 7\mathbf{j} + 2\mathbf{k}$

Hence $AB = |\overrightarrow{AB}| = \sqrt{(1)^2 + (-1)^2 + (1)^2} = \sqrt{3}$

          $BC = |\overrightarrow{BC}| = \sqrt{(-4)^2 + (8)^2 + (-3)^2} = \sqrt{89}$

          $CA = |\overrightarrow{CA}| = \sqrt{(3)^2 + (-7)^2 + (2)^2} = \sqrt{62}$

The perimeter $= \sqrt{3} + \sqrt{89} + \sqrt{62} = 19.0$ units (3 s.f.)

The coordinate axes are not drawn in this diagram as they complicate the diagram when two or more points are illustrated. The origin should always be included, however, as it provides a reference point.

$\overrightarrow{AB}$ is 'clockwise', $\overrightarrow{AO}$ and $\overrightarrow{OB}$ are 'anticlockwise'.

Two-dimensional problems can be solved by using the same principles as for three-dimensional cases but the working tends to be easier because it involves fewer terms.

Examples 13a cont.

3 Given that $\mathbf{p} = \mathbf{i} + 3\mathbf{j}$, $\mathbf{q} = 4\mathbf{i} - 2\mathbf{j}$, $\overrightarrow{OA} = 2\mathbf{p}$ and $\overrightarrow{OB} = 3\mathbf{q}$, find

  (a) $|\overrightarrow{OA}|$

  (b) $|\overrightarrow{OB}|$

  (c) $|\overrightarrow{AB}|$

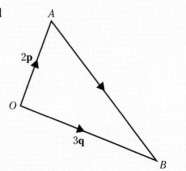

(a) $|\overrightarrow{OA}| = 2|\mathbf{i} + 3\mathbf{j}| = 2\sqrt{1^2 + 3^2} = 2\sqrt{10} = 6.32$ units (3 s.f.)

(b) $|\overrightarrow{OB}| = 3|4\mathbf{i} - 2\mathbf{j}| = 3\sqrt{4^2 + (-2)^2} = 6\sqrt{5} = 13.4$ units (3 s.f.)

(c) $\overrightarrow{AB} = \overrightarrow{AO} + \overrightarrow{OB} = \overrightarrow{OB} - \overrightarrow{OA} = 3(4\mathbf{i} - 2\mathbf{j}) - 2(\mathbf{i} + 3\mathbf{j}) = 10\mathbf{i} - 12\mathbf{j}$

$|\overrightarrow{AB}| = \sqrt{10^2 + 12^2} = 2\sqrt{61} = 15.6$ units (3 s.f.)

Exercise 13a

1 Write down, in the form $a\mathbf{i} + b\mathbf{j} + c\mathbf{k}$, the vector represented by $\overrightarrow{OP}$ when P is a point with coordinates

(a) $(3, 6, 4)$ (b) $(1, -2, -7)$

(c) $(1, 0, -3)$

2 $\overrightarrow{OP}$ represents a vector $\mathbf{r}$. Write down the coordinates of P when

(a) $\mathbf{r} = 5\mathbf{i} - 7\mathbf{j} + 2\mathbf{k}$

(b) $\mathbf{r} = \mathbf{i} + 4\mathbf{j}$

(c) $\mathbf{r} = \mathbf{j} - \mathbf{k}$

3 Find the length of the line OP when P is the point

(a) $(2, 1, 4)$ (b) $(3, 0, 4)$

(c) $(-2, -2, 1)$

4 Find the modulus of the vector $\mathbf{V}$ when

(a) $\mathbf{V} = 2\mathbf{i} - 4\mathbf{j} + 4\mathbf{k}$

(b) $\mathbf{V} = 6\mathbf{i} + 2\mathbf{j} - 3\mathbf{k}$

(c) $\mathbf{V} = 11\mathbf{i} - 7\mathbf{j} - 6\mathbf{k}$

5 If $\mathbf{a} = \mathbf{i} + \mathbf{j} + \mathbf{k}$, $\mathbf{b} = 2\mathbf{i} - \mathbf{j} + 3\mathbf{k}$, $\mathbf{c} = -\mathbf{i} + 3\mathbf{j} - \mathbf{k}$ find

(a) $\mathbf{a} + \mathbf{b}$ (b) $\mathbf{a} - \mathbf{c}$

(c) $\mathbf{a} + \mathbf{b} + \mathbf{c}$ (d) $\mathbf{a} - 2\mathbf{b} + 3\mathbf{c}$

In questions **6** to **8**, $\overrightarrow{OA} = \mathbf{a} = 4\mathbf{i} - 12\mathbf{j}$ and $\overrightarrow{OB} = \mathbf{b} = \mathbf{i} + 6\mathbf{j}$

6 Which of the following vectors are parallel to $\mathbf{a}$?

(a) $\mathbf{i} + 3\mathbf{j}$ (b) $-\mathbf{i} + 3\mathbf{j}$

(c) $12\mathbf{i} - 4\mathbf{j}$ (d) $-4\mathbf{i} + 12\mathbf{j}$

(e) $\mathbf{i} - 3\mathbf{j}$

7 Which of the following vectors are equal to $\mathbf{b}$?

(a) $2\mathbf{i} + 12\mathbf{j}$

(b) $-\mathbf{i} - 6\mathbf{j}$

(c) $\overrightarrow{AE}$ if E is $(5, -6)$

(d) $\overrightarrow{AF}$ if F is $(6, 0)$

8 When $\overrightarrow{OD} = \lambda\overrightarrow{OA}$, find the value of λ for which $\overrightarrow{OD} + \overrightarrow{OB}$ is parallel to the x-axis.

9 Which of the following vectors are parallel to $3\mathbf{i} - \mathbf{j} - 2\mathbf{k}$?

(a) $6\mathbf{i} - 3\mathbf{j} - 4\mathbf{k}$ (b) $-9\mathbf{i} + 3\mathbf{j} + 6\mathbf{k}$

(c) $-3\mathbf{i} - \mathbf{j} - 2\mathbf{k}$ (d) $-2(3\mathbf{i} + \mathbf{j} + 2\mathbf{k})$

(e) $\frac{3}{2}\mathbf{i} - \frac{1}{2}\mathbf{j} - \mathbf{k}$ (f) $-\mathbf{i} + \frac{1}{3}\mathbf{j} + \frac{2}{3}\mathbf{k}$

10 Given that $\mathbf{a} = 4\mathbf{i} + \mathbf{j} - 6\mathbf{k}$, state whether each of the following vectors is parallel or equal to $\mathbf{a}$ or neither.

(a) $8\mathbf{i} + 2\mathbf{j} - 10\mathbf{k}$

(b) $-4\mathbf{i} - \mathbf{j} + 6\mathbf{k}$

(c) $2(2\mathbf{i} + \frac{1}{2}\mathbf{j} - 3\mathbf{k})$

11 The triangle ABC has its vertices at the points $A(-1, 3, 0)$, $B(-3, 0, 7)$, $C(-1, 2, 3)$. Find in the form $a\mathbf{i} + b\mathbf{j} + c\mathbf{k}$ the vectors representing

(a) $\overrightarrow{AB}$ (b) $\overrightarrow{AC}$ (c) $\overrightarrow{CB}$

12 Find the lengths of the sides of the triangles described in question **11**.

13 Find $|\mathbf{a} - \mathbf{b}|$ where $\mathbf{a} = \mathbf{i} - \mathbf{j} + 2\mathbf{k}$, $\mathbf{b} = 2\mathbf{i} - \mathbf{j}$

14 A, B, C and D are the points $(0, 0, 2)$, $(-1, 3, 2)$, $(1, 0, 4)$ and $(-1, 2, -2)$ respectively. Find the vectors representing $\overrightarrow{AB}$, $\overrightarrow{BD}$, $\overrightarrow{CD}$, $\overrightarrow{AD}$.

15

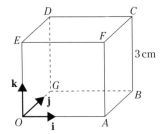

OABCDEFG is a cube of side 3 cm. *O* is the origin and unit vectors **i**, **j**, and **k** are parallel to *OA*, *OG* and *OE* respectively.

Express each of the vectors $\overrightarrow{OB}$, $\overrightarrow{OC}$ and $\overrightarrow{GF}$ in terms of **i**, **j**, and **k**.

16

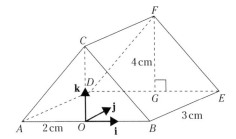

ABCDEF is a triangular prism with the dimensions shown in the diagram. *O* is the origin and the midpoint of *AB*. *G* is the midpoint of *DE*. The unit vectors **i**, **j**, and **k** are parallel to *OB*, *OG* and *OC* respectively.

Express each of the vectors $\overrightarrow{OE}$, $\overrightarrow{EF}$ and $\overrightarrow{FA}$ in terms of **i**, **j**, and **k**.

17

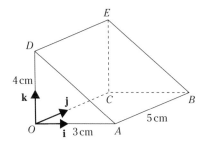

OABCDE is a triangular prism with the dimensions shown on the diagram. The unit vectors **i**, **j**, and **k** are parallel to *OA*, *OC* and *OD* respectively.

Express each of the vectors $\overrightarrow{OE}$, $\overrightarrow{EB}$ and $\overrightarrow{EA}$ in terms of **i**, **j**, and **k**.

18

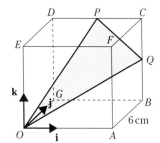

OABCDE is a cube with a side of 6 cm. *P* and *Q* are the midpoints of *DC* and *BC* respectively. The unit vectors **i**, **j**, and **k** are parallel to *OA*, *OG* and *OE* respectively.

(a) Express each of the vectors $\overrightarrow{OP}$, $\overrightarrow{OQ}$ and $\overrightarrow{PQ}$ in terms of **i**, **j** and **k**.

(b) Calculate the perimeter of triangle *OPQ*.

FINDING A UNIT VECTOR

The vector **v** = 6**i** + 2**j** + 3**k** represents $\overrightarrow{OP}$.

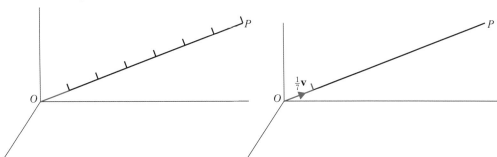

$|\mathbf{v}|$ is $\sqrt{6^2 + 2^2 + 3^2}$, i.e. *OP* is 7 units long.

Therefore $\frac{1}{7}\mathbf{v}$ is a vector of unit magnitude and is denoted by $\hat{\mathbf{v}}$

i.e. $\hat{\mathbf{v}} = \dfrac{\mathbf{v}}{|\mathbf{v}|}$

In general   **a unit vector in the direction of v is given by** $\dfrac{\mathbf{v}}{|\mathbf{v}|}$

Example 13b

Find the vector $\overrightarrow{OP}$ given that $\overrightarrow{OP}$ is of length 5 units and is in the direction of the vector $2\mathbf{i} - \mathbf{j} + 4\mathbf{k}$.

The unit vector in the direction of $2\mathbf{i} - \mathbf{j} + 4\mathbf{k} = \dfrac{2\mathbf{i} - \mathbf{j} + 4\mathbf{k}}{\sqrt{4 + 1 + 16}} = \dfrac{1}{\sqrt{21}}(2\mathbf{i} - \mathbf{j} + 4\mathbf{k})$

$\therefore \qquad \overrightarrow{OP} = \dfrac{5}{\sqrt{21}}(2\mathbf{i} - \mathbf{j} + 4\mathbf{k})$

Exercise 13b

1 Find a unit vector in the direction of each of the vectors.

    (a) $2\mathbf{i} + 2\mathbf{j} - \mathbf{k}$ (b) $6\mathbf{i} - 2\mathbf{j} - 3\mathbf{k}$

    (c) $3\mathbf{i} + 4\mathbf{k}$ (d) $\mathbf{i} + 8\mathbf{j} + 4\mathbf{k}$

    (e) $3\mathbf{i} + 4\mathbf{j}$

2 Relative to an origin O, the position vectors of points A and B are $\overrightarrow{OA} = 3\mathbf{i} + 2\mathbf{j} - \mathbf{k}$ and $\overrightarrow{OB} = \mathbf{i} - \mathbf{j} + 2\mathbf{k}$ respectively.

Find the unit vector in the direction of $\overrightarrow{AB}$.

3 Relative to an origin O, the position vectors of points A and B are $\overrightarrow{OA} = -\mathbf{i} + 2\mathbf{j} + \mathbf{k}$ and $\overrightarrow{OB} = \mathbf{i} - a\mathbf{j} + \mathbf{k}$ respectively.

    (a) Find the unit vector in the direction of $\overrightarrow{AB}$ in terms of a.

    (b) Find the value of a for which $|\overrightarrow{AB}| = 2$

4 Relative to an origin O, the position vectors of points A and B are $\overrightarrow{OA} = \mathbf{i} + 2\mathbf{j} + 3\mathbf{k}$ and $\overrightarrow{OB} = \mathbf{i} - 3\mathbf{j} + b\mathbf{k}$ respectively.

Find the values of b for which $|\overrightarrow{AB}| = 5$

THE ANGLE BETWEEN TWO VECTORS

There are two angles between two lines, i.e. α and $180° - \alpha$.

The angle between two vectors, however, is defined as the angle between their directions when the lines representing them *both converge* or *both diverge* (see diagrams a and b).

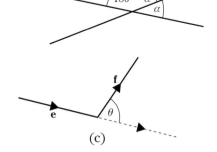

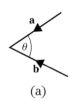

(a)   (b)   (c)

In some cases one of the lines may have to be extended in order to mark the correct angle (see diagram c).

THE SCALAR PRODUCT

The scalar product of two vectors $\mathbf{a}$ and $\mathbf{b}$ is denoted by $\mathbf{a} . \mathbf{b}$ and defined as $ab \cos \theta$ where θ is the angle between $\mathbf{a}$ and $\mathbf{b}$.

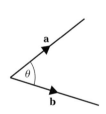

i.e. $\mathbf{a} . \mathbf{b} = ab \cos \theta$

As $ab \cos \theta = ba \cos \theta$ it follows that $\mathbf{a} . \mathbf{b} = \mathbf{b} . \mathbf{a}$

Parallel vectors

If **a** and **b** are parallel then

either $\mathbf{a} \cdot \mathbf{b} = ab \cos 0$ or $\mathbf{a} \cdot \mathbf{b} = ab \cos \pi$

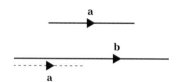   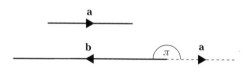

i.e.   **for like parallel vectors** $\mathbf{a} \cdot \mathbf{b} = ab$

and   **for unlike parallel vectors** $\mathbf{a} \cdot \mathbf{b} = -ab$

In the special case when $\mathbf{a} = \mathbf{b}$

$\mathbf{a} \cdot \mathbf{b} = \mathbf{a} \cdot \mathbf{a} = a^2$ (sometimes $\mathbf{a} \cdot \mathbf{a}$ is written $\mathbf{a}^2$)

For the unit vectors **i**, **j** and **k**,

$\mathbf{i} \cdot \mathbf{i} = \mathbf{j} \cdot \mathbf{j} = \mathbf{k} \cdot \mathbf{k} = 1$

Perpendicular vectors

If **a** and **b** are perpendicular then $\theta = \frac{1}{2}\pi$, $\Rightarrow$ $\mathbf{a} \cdot \mathbf{b} = ab \cos \frac{1}{2}\pi = 0$

i.e.   **for perpendicular vectors a . b = 0**

For the unit vectors **i**, **j** and **k**,

$\mathbf{i} \cdot \mathbf{j} = \mathbf{j} \cdot \mathbf{k} = \mathbf{k} \cdot \mathbf{i} = 0$

CALCULATING a . b IN CARTESIAN FORM

When $\mathbf{a} = x_1\mathbf{i} + y_1\mathbf{j} + z_1\mathbf{k}$ and $\mathbf{b} = x_2\mathbf{i} + y_2\mathbf{j} + z_2\mathbf{k}$

$\mathbf{a} \cdot \mathbf{b} = (x_1x_2 + y_1y_2 + z_1z_2)$

i.e. $(x_1\mathbf{i} + y_1\mathbf{j} + z_1\mathbf{k}) \cdot (x_2\mathbf{i} + y_2\mathbf{j} + z_2\mathbf{k}) = x_1x_2 + y_1y_2 + z_1z_2$

So $(2\mathbf{i} - 3\mathbf{j} + 4\mathbf{k}) \cdot (\mathbf{i} + 3\mathbf{j} - 2\mathbf{k}) = (2)(1) + (-3)(3) + (4)(-2) = -15$

Examples 13c

1 Find the scalar product of $\mathbf{a} = 2\mathbf{i} - 3\mathbf{j} + 5\mathbf{k}$ and $\mathbf{b} = \mathbf{i} - 3\mathbf{j} + \mathbf{k}$ and hence find the angle between **a** and **b**.

$\mathbf{a} \cdot \mathbf{b} = (2)(1) + (-3)(-3) + (5)(1) = 16$

But $\mathbf{a} \cdot \mathbf{b} = |\mathbf{a}| \, |\mathbf{b}| \cos \theta$

$|\mathbf{a}| = \sqrt{4 + 9 + 25} = \sqrt{38}$ and $|\mathbf{b}| = \sqrt{1 + 9 + 1} = \sqrt{11}$

Hence $\cos \theta = \dfrac{\mathbf{a} \cdot \mathbf{b}}{|\mathbf{a}| \, |\mathbf{b}|} = \dfrac{16}{\sqrt{11}\sqrt{38}} = 0.7825\ldots$

$\therefore$ $\theta = 38.5°$ (1 d.p.)

2 Relative to an origin O, the position vectors of points A and B are $\overrightarrow{OA} = \mathbf{i} + 2\mathbf{j} + 3\mathbf{k}$ and $\overrightarrow{OB} = \mathbf{i} - 3\mathbf{j} + 2\mathbf{k}$ respectively. State whether angle AOB is acute, obtuse or a right-angle.

$$\overrightarrow{OA} \cdot \overrightarrow{OB} = 1 - 6 + 6 = 1$$

$$\therefore \quad |\overrightarrow{OA}| \times |\overrightarrow{OB}| \times \cos \angle AOB = 1$$

so $\cos \angle AOB = \dfrac{1}{|\overrightarrow{OA}| \, |\overrightarrow{OB}|}$ which is positive.

$\therefore \quad \angle AOB$ is acute.

3 Find the value of a for which the vectors $\begin{pmatrix} 2 \\ -1 \\ 1 \end{pmatrix}$ and $\begin{pmatrix} 2 \\ a \\ -2 \end{pmatrix}$ are perpendicular.

$$\begin{pmatrix} 2 \\ -1 \\ 1 \end{pmatrix} \cdot \begin{pmatrix} 2 \\ a \\ -2 \end{pmatrix} = 4 - a - 2$$

The vectors are perpendicular when $4 - a - 2 = 0$

$\therefore \quad a = 2$

Exercise 13c

1 Calculate $\mathbf{a} \cdot \mathbf{b}$ if

(a) $\mathbf{a} = 2\mathbf{i} - 4\mathbf{j} + 5\mathbf{k}$, $\mathbf{b} = \mathbf{i} + 3\mathbf{j} + 8\mathbf{k}$

(b) $\mathbf{a} = 3\mathbf{i} - 7\mathbf{j} + 2\mathbf{k}$, $\mathbf{b} = 5\mathbf{i} + \mathbf{j} - 4\mathbf{k}$

(c) $\mathbf{a} = 2\mathbf{i} - 3\mathbf{j} + 6\mathbf{k}$, $\mathbf{b} = \mathbf{i} + \mathbf{j}$

What conclusion can you draw in part **b**?

2 Find $\mathbf{p} \cdot \mathbf{q}$ and the cosine of the angle between $\mathbf{p}$ and $\mathbf{q}$ if

(a) $\mathbf{p} = 2\mathbf{i} + 4\mathbf{j} + \mathbf{k}$, $\mathbf{q} = \mathbf{i} + \mathbf{j} + \mathbf{k}$

(b) $\mathbf{p} = -\mathbf{i} + 3\mathbf{j} - 2\mathbf{k}$, $\mathbf{q} = \mathbf{i} + \mathbf{j} - 6\mathbf{k}$

(c) $\mathbf{p} = -2\mathbf{i} + 5\mathbf{j}$, $\mathbf{q} = \mathbf{i} + \mathbf{j}$

(d) $\mathbf{p} = 2\mathbf{i} + \mathbf{j}$, $\mathbf{q} = \mathbf{j} - 2\mathbf{k}$

3 The cosine of the angle between two vectors $\mathbf{v}_1$ and $\mathbf{v}_2$ is $\frac{4}{21}$

If $\mathbf{v}_1 = 6\mathbf{i} + 3\mathbf{j} - 2\mathbf{k}$ and $\mathbf{v}_2 = -2\mathbf{i} + \lambda\mathbf{j} - 4\mathbf{k}$, find the positive value of λ

4 In a triangle ABC, $\overrightarrow{AB} = \mathbf{i} + 2\mathbf{j} + 3\mathbf{k}$ and $\overrightarrow{BC} = -\mathbf{i} + 4\mathbf{j}$

Find the cosine of angle ABC.

Find the vector $\overrightarrow{AC}$ and use it to calculate the angle BAC.

5 Show that $\mathbf{i} + 7\mathbf{j} + 3\mathbf{k}$ is perpendicular to both $\mathbf{i} - \mathbf{j} + 2\mathbf{k}$ and $2\mathbf{i} + \mathbf{j} - 3\mathbf{k}$.

6 Show that $13\mathbf{i} + 23\mathbf{j} + 7\mathbf{k}$ is perpendicular to both $2\mathbf{i} + \mathbf{j} - 7\mathbf{k}$ and $3\mathbf{i} - 2\mathbf{j} + \mathbf{k}$.

7 The magnitudes of two vectors $\mathbf{p}$ and $\mathbf{q}$ are 5 and 4 units respectively.

The angle between $\mathbf{p}$ and $\mathbf{q}$ is $30°$.

Find (a) $\mathbf{p} \cdot \mathbf{q}$

        (b) the magnitude of the vector $\mathbf{p} - \mathbf{q}$

8 Calculate the acute angle between the vectors $\begin{pmatrix} 2 \\ -1 \\ 3 \end{pmatrix}$ and $\begin{pmatrix} 0 \\ -1 \\ -1 \end{pmatrix}$

9   The diagram shows a cube where the length of each edge is 4 cm.

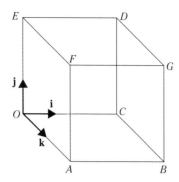

(a)   Express in terms of **i**, **j** and **k** the vectors
    (i) $\overrightarrow{AE}$
    (ii) $\overrightarrow{AG}$

(b)   Given that *H* is the midpoint of *AB*, find the angles of the triangle *AEH*.

10 *OABCDEFG* is a cube where the length of each edge is 4 cm.

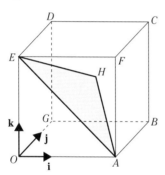

H is the point where the diagonals of the face *BCDG* meet.

(a)   Express in terms of **i**, **j** and **k** the vectors
    (i) $\overrightarrow{AE}$
    (ii) $\overrightarrow{AH}$

(b)   Find the angle *EHA*.

11 In triangle *OAB*, *O* is the origin,
    $\overrightarrow{OA} = 4\mathbf{i} - 3\mathbf{j} + 4\mathbf{k}$ and $\overrightarrow{OB} = \mathbf{i} + 6\mathbf{j} - 2\mathbf{k}$

(a)   Show that triangle *OAB* is isosceles.

(b)   Find angle *AOB* correct to the nearest degree.

(c)   Hence or otherwise find the area of triangle *OAB*.

12 The points *A*, *B*, *C* have position vectors
    $\mathbf{a} = 2\mathbf{i} + \mathbf{j} - \mathbf{k}$, $\mathbf{b} = 3\mathbf{i} + 4\mathbf{j} - 2\mathbf{k}$,
    $\mathbf{c} = 5\mathbf{i} - \mathbf{j} + 2\mathbf{k}$

    respectively, relative to a fixed origin *O*.

(a)   Calculate the scalar product
    $(\mathbf{a} - \mathbf{b}) . (\mathbf{c} - \mathbf{b})$ and hence calculate the angle *ABC*.

(b)   Given that *ABCD* is a parallelogram:
    (i)   determine the position vector of *D*;
    (ii) calculate the area of *ABCD*.

(c)   The point *E* lies on *BA* produced so that $\overrightarrow{BE} = \overrightarrow{3BA}$

    Write down the position vector of *E*.

    The line *CE* cuts the line *AD* at *X*. Find the position vector of *X*.

13

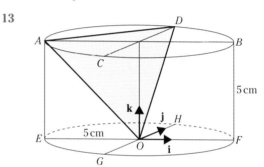

The diagram shows a right circular cylinder. The height of the cylinder is 5 cm and the radius of the base is 5 cm. *AB* and *EF* are parallel diameters of the circular ends.
CD and *GH* are parallel diameters and are perpendicular to *AB* and *EF* respectively.
The unit vectors **i**, **j**, and **k** are parallel to *OA*, *OG* and *OE* respectively.

Calculate the size of angle *AOD*.

14

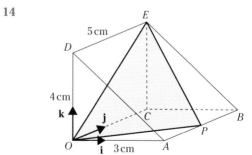

OABCDE is a triangular prism with the dimensions shown on the diagram. *P* is the midpoint of *AB*. The unit vectors **i**, **j**, and **k** are parallel to *OA*, *OC* and *OD* respectively.

(a)   Calculate the size of the angle *EOP*.

(b)   Calculate the area of triangle *OEP*.

Summary 3

TRIGONOMETRIC RATIOS OF 30°, 60°, 45°

	30°	60°	45°
sin	$\dfrac{1}{2}$	$\dfrac{\sqrt{3}}{2}$	$\dfrac{1}{\sqrt{2}}$
cos	$\dfrac{\sqrt{3}}{2}$	$\dfrac{1}{2}$	$\dfrac{1}{\sqrt{2}}$
tan	$\dfrac{1}{\sqrt{3}}$	$\sqrt{3}$	1

TRIGONOMETRIC FUNCTIONS

$$f(x) = \sin x$$

The sine function, $f(x) = \sin x$

is defined for all values of x

is periodic with a period 2π

has a maximum value of 1 when $x = \left(2n + \frac{1}{2}\right)\pi$

and a minimum value of -1 when $x = \left(2n + \frac{3}{2}\right)\pi$

is zero when $x = n\pi$

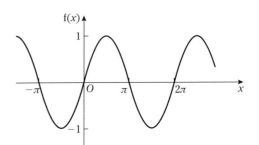

$$f(x) = \cos x$$

The cosine function, $f(x) = \cos x$

is defined for all values of x

is periodic with a period 2π

has a maximum value of 1 when $x = 2n\pi$

and a minimum value of -1 when $x = (2n + 1)\pi$

is zero when $x = \frac{1}{2}(2n + 1)\pi$

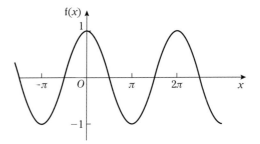

$$f(x) = \tan x$$

The tangent function, $y = \tan x$

is undefined for some values of x,

these values being all odd multiples of $\frac{1}{2}\pi$,

is periodic with a period π.

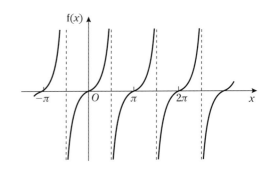

INVERSE TRIGONOMETRIC FUNCTIONS

The inverse trigonometric functions are $\sin^{-1} x$, $\cos^{-1} x$, $\tan^{-1} x$.

$\sin^{-1} x$ means 'the angle in the range $-\frac{1}{2}\pi \leqslant \theta \leqslant \frac{1}{2}\pi$ whose sine is x'.

$\cos^{-1} x$ means 'the angle in the range $0 \leqslant \theta \leqslant \pi$ whose cosine is x'.

$\tan^{-1} x$ means 'the angle in the range $-\frac{1}{2}\pi \leqslant \theta \leqslant \frac{1}{2}\pi$ whose tangent is x'.

Trigonometric identities

$\tan \theta \equiv \dfrac{\sin \theta}{\cos \theta}$

$\sin^2 \theta + \cos^2 \theta \equiv 1$

ARITHMETIC PROGRESSIONS

In an arithmetic progression, each term differs from the preceding term by a constant (called the common difference).

An AP with first term a, common difference d and n terms is

$$a, a + d, a + 2d, \dots , \{a + (n - 1)d\}$$

The sum of the first n terms is given by

$$S_n = \tfrac{1}{2}n(a + l) \text{ where } l \text{ is the last term}$$
$$= \tfrac{1}{2}n\{2a + (n - 1)d\}$$

GEOMETRIC PROGRESSIONS

In a geometric progression each term is a constant multiple of the preceding term. This multiple is called the common ratio.

A GP with first term a, common ratio r and n terms is

$$a, ar, ar^2, \dots , ar^{n-1}$$

The sum of the first n terms is given by

$$S_n = \frac{a(1 - r^n)}{1 - r}$$

The sum to infinity is given by $S = \dfrac{a}{1 - r}$ provided that $-1 < r < 1$

The binomial theorem

If n is a positive integer then $(a + b)^n$ can be expanded as a finite series, where

$$(a + b)^n = a^n + na^{n-1} b + \binom{n}{2}a^{n-2} b^2 + \binom{n}{3} a^{n-3} b^3 + \dots + b^n$$

and where

$$\binom{n}{r} = \frac{n(n - 1)(n - 2) \dots (n - r + 1)}{r!}$$

In particular, $(1 + x)^n = 1 + nx + \binom{n}{2}x^2 + \binom{n}{3}x^3 + \dots + x^n$

INTEGRATION

Standard integrals

Function	Integral
x^n	$\dfrac{1}{n + 1}x^{n+1}, (n \neq -1)$
$(ax + b)^n$	$\dfrac{1}{a(n + 1)}(ax + b)^{n+1}, n \neq -1$

Area

The area bounded by the lines $x = a$ and $x = b$ and the curve $y = f(x)$ is given by

$$\text{Area} = \int_a^b f(x)\, dx$$

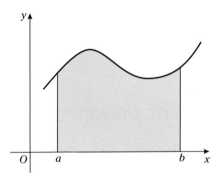

Volume of revolution

The volume formed by rotating the area bounded by the curve $y = f(x)$, the lines $x = a$ and $x = b$ and the x-axis is given by

$$\text{Volume} = \int_a^b \pi y^2\, dx$$

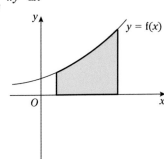

VECTORS

A vector is a quantity with both magnitude and direction and can be represented by a line segment.

If lines representing several vectors are drawn 'head to tail' in order, then the line (in the opposite sense) which completes a closed polygon represents the sum of the vectors (or the resultant vector).

A position vector has a fixed location in space.

Cartesian unit vectors

$\mathbf{i}, \mathbf{j}$ and $\mathbf{k}$ are unit vectors in the directions of Ox, Oy and Oz respectively.

Any vector can be given in the form $a\mathbf{i} + b\mathbf{j} + c\mathbf{k}$.

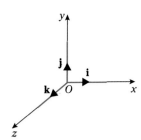

$$(a_1\mathbf{i} + b_1\mathbf{j} + c_1\mathbf{k}) \pm (a_2\mathbf{i} + b_2\mathbf{j} + c_2\mathbf{k}) = (a_1 \pm a_2)\mathbf{i} + (b_1 \pm b_2)\mathbf{j} + (c_1 \pm c_2)\mathbf{k}$$

$$|a\mathbf{i} + b\mathbf{j} + c\mathbf{k}| = \sqrt{a^2 + b^2 + c^2}$$

For two vectors $\mathbf{v}_1 = a_1\mathbf{i} + b_1\mathbf{j} + c_1\mathbf{k}$ and $\mathbf{v}_2 = a_2\mathbf{i} + b_2\mathbf{j} + c_2\mathbf{k}$

$\mathbf{v}_1$ and $\mathbf{v}_2$ are parallel if $\mathbf{v}_1 = \lambda\mathbf{v}_2$

i.e. $\quad a_1 = \lambda a_2, b_1 = \lambda b_2, c_1 = \lambda c_2$

$\mathbf{v}_1$ and $\mathbf{v}_2$ are equal if $a_1 = a_2, b_1 = b_2, c_1 = c_2$

The scalar product of two vectors

If θ is the angle between two vectors $\mathbf{a}$ and $\mathbf{b}$ then

$$\mathbf{a} \cdot \mathbf{b} = |\mathbf{a}|\,|\mathbf{b}| \cos \theta$$

$\mathbf{a}$ and $\mathbf{b}$ are perpendicular $\Rightarrow \mathbf{a} \cdot \mathbf{b} = 0$

If $\mathbf{a} = x_1\mathbf{i} + y_1\mathbf{j} + z_1\mathbf{k}$ and $\mathbf{b} = x_2\mathbf{i} + y_2\mathbf{j} + z_2\mathbf{k}$ then

$$\mathbf{a} \cdot \mathbf{b} = x_1 x_2 + y_1 y_2 + z_1 z_2$$

Summary exercise 3

1 Find the value of the coefficient of x^2 in the expansion of $\left(\dfrac{x}{2} + \dfrac{2}{x}\right)^6$. [3]

Cambridge, Paper 1 Q1 N08

2 Prove the identity

$$\frac{1 + \sin x}{\cos x} + \frac{\cos x}{1 + \sin x} \equiv \frac{2}{\cos x} \quad [4]$$

Cambridge, Paper 1 Q2 N08

3 The first term of an arithmetic progression is 6 and the fifth term is 12. The progression has n terms and the sum of all the terms is 90. Find the value of n. [4]

Cambridge, Paper 1 Q3 N08

4

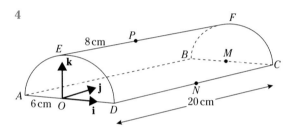

The diagram shows a semicircular prism with a horizontal rectangular base $ABCD$. The vertical ends AED and BFC are semicircles of radius 6 cm. The length of the prism is 20 cm. The mid-point of AD is the origin O, the mid-point of BC is M and the mid-point of DC is N. The points E and F are the highest points of the semicircular ends of the prism. The point P lies on EF such that $EP = 8$ cm

Unit vectors $\mathbf{i}$, $\mathbf{j}$ and $\mathbf{k}$ are parallel to OD, OM and OE respectively.

(i) Express each of the vectors $\overrightarrow{PA}$ and $\overrightarrow{PN}$ in terms of $\mathbf{i}$, $\mathbf{j}$ and $\mathbf{k}$. [3]

(ii) Use a scalar product to calculate angle APN. [4]

Cambridge, Paper 1 Q4 N08

5 Solve the equation $3 \tan(2x + 15°) = 4$ for $0° \leqslant x \leqslant 180°$ [4]

Cambridge, Paper 11 Q1 N09

6 The equation of a curve is $y = 3 \cos 2x$. The equation of a line is $x + 2y = \pi$. On the same diagram, sketch the curve and the line for $0 \leqslant x \leqslant \pi$ [4]

Cambridge, Paper 11 Q2 N09

7

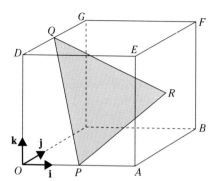

The diagram shows a cube $OABCDEFG$ in which the length of each side is 4 units. The unit vectors $\mathbf{i}$, $\mathbf{j}$ and $\mathbf{k}$ are parallel to $\overrightarrow{OA}$, $\overrightarrow{OC}$ and $\overrightarrow{OD}$ respectively. The mid-points of OA and DG are P and Q respectively and R is the centre of the square face $ABFE$.

(i) Express each of the vectors $\overrightarrow{PR}$ and $\overrightarrow{PQ}$ in terms of $\mathbf{i}$, $\mathbf{j}$ and $\mathbf{k}$. [3]

(ii) Use a scalar product to find angle QPR. [4]

(iii) Find the perimeter of triangle PQR, giving your answer correct to 1 decimal place. [3]

Cambridge, Paper 1 Q10 N07

8

The diagram shows part of the curve $y = x + \dfrac{4}{x}$ which has a minimum point at M. The line $y = 5$ intersects the curve at the points A and B.

(i) Find the coordinates of A, B and M. [5]

(ii) Find the volume obtained when the shaded region is rotated through 360° about the x-axis. [6]

Cambridge, Paper 13 Q9 J10

9 (i) Find the sum to infinity of the geometric progression with first three terms 0.5, 0.5^3 and 0.5^5 [3]

(ii) The first two terms in an arithmetic progression are 5 and 9. The last term in the progression is the only term which is greater than 200. Find the sum of all the terms in the progression.　　[4]

Cambridge, Paper 1 Q7 J09

10 (i) Show that the equation
$2 \tan^2 \theta \cos \theta = 3$ can be written in the form $2 \cos^2 \theta + 3 \cos \theta - 2 = 0$　　[2]

(ii) Hence solve the equation
$2 \tan^2 \theta \cos \theta = 3$, for $0° \leq \theta \leq 360°$　　[3]

Cambridge, Paper 1 Q2 J08

11 The function f is defined by
$f(x) = a + b \cos 2x$, for $0 \leq x \leq \pi$. It is given that $f(0) = -1$ and $f\left(\frac{1}{2}\pi\right) = 7$

(i) Find the values of a and b.　　[3]

(ii) Find the x-coordinates of the points where the curve $y = f(x)$ intersects the x-axis.　　[3]

(iii) Sketch the graph of $y = f(x)$.　　[2]

Cambridge, Paper 1 Q8 J07

12

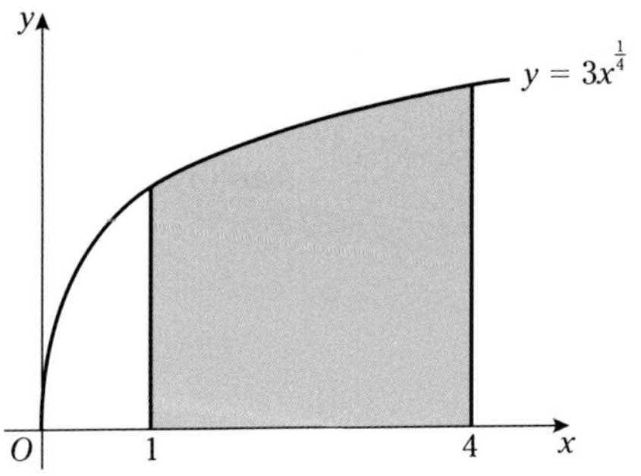

The diagram shows the curve $y = 3x^{\frac{1}{4}}$. The shaded region is bounded by the curve, the x-axis and the lines $x = 1$ and $x = 4$. Find the volume of the solid obtained when this shaded region is rotated completely about the x-axis, giving your answer in terms of π　　[4]

Cambridge, Paper 1 Q2 J07

13 Prove the identity
$$\frac{1 - \tan^2 x}{1 + \tan^2 x} = 1 - 2 \sin^2 x \qquad [4]$$

Cambridge, Paper 1 Q3 J07

14 Relative to an origin O, the position vectors of the points A, B, C and D are given by

$$\overrightarrow{OA} = \begin{pmatrix} 1 \\ 3 \\ -1 \end{pmatrix}, \overrightarrow{OB} = \begin{pmatrix} 3 \\ -1 \\ 3 \end{pmatrix}, \overrightarrow{OC} = \begin{pmatrix} 4 \\ 2 \\ p \end{pmatrix} \text{ and}$$

$$\overrightarrow{OD} = \begin{pmatrix} -1 \\ 0 \\ q \end{pmatrix}$$

where p and q are constants. Find

(i) the unit vector in the direction of $\overrightarrow{AB}$　　[3]

(ii) the value of p for which angle $AOC = 90°$　　[3]

(iii) the values of q for which the length of $\overrightarrow{AD}$ is 7 units.　　[4]

Cambridge, Paper 1 Q9 J04

List of formulae

Algebra

For the quadratic equation $ax^2 + bx + c = 0$:

$$x = \frac{-b \pm \sqrt{b^2 - 4ac}}{2a}$$

For an arithmetic series:

$$u_n = a + (n-1)d, \qquad S_n = \tfrac{1}{2}n(a + l) = \tfrac{1}{2}n\,\{2a + (n-1)d\}$$

For a geometric series:

$$u_n = ar^{n-1}, \qquad S_n = \frac{a(1 - r^n)}{1 - r} \ \ (r \neq 1), \qquad S_\infty = \frac{a}{1 - r} \qquad (|r| < 1)$$

Binomial expansion:

$$(a + b)^n = a^n + \binom{n}{1} a^{n-1} b + \binom{n}{2} a^{n-2} b^2 + \binom{n}{3} a^{n-3} b^3 + \ldots + b^n, \text{ where } n \text{ is a positive integer}$$

$$\text{and } \binom{n}{r} = \frac{n!}{r!(n - r)!}$$

$$(1 + x)^n = 1 + nx + \frac{n(n - 1)}{2!} x^2 + \frac{n(n - 1)(n - 2)}{3!} x^3 \ldots, \text{ where } n \text{ is rational and } |x| < 1$$

Trigonometry

Arc length of circle $= r\theta$ (θ in radians)

Area of sector of circle $= \tfrac{1}{2} r^2 \theta$ (θ in radians)

$$\tan\theta \equiv \frac{\sin\theta}{\cos\theta}$$

$$\cos^2\theta + \sin^2\theta \equiv 1, \qquad 1 + \tan^2\theta \equiv \sec^2\theta, \qquad \cot^2\theta + 1 \equiv \operatorname{cosec}^2\theta$$

$$\sin(A \pm B) \equiv \sin A \cos B \pm \cos A \sin B$$

$$\cos(A \pm B) \equiv \cos A \cos B \mp \sin A \sin B$$

$$\tan(A \pm B) \equiv \frac{\tan A \pm \tan B}{1 \mp \tan A \tan B}$$

$$\sin 2A \equiv 2 \sin A \cos A$$

$$\cos 2A \equiv \cos^2 A - \sin^2 A \equiv 2\cos^2 A - 1 \equiv 1 - 2\sin^2 A$$

$$\tan 2A = \frac{2 \tan A}{1 - \tan^2 A}$$

Principal values:

$$-\tfrac{1}{2}\pi \leqslant \sin^{-1} x \leqslant \tfrac{1}{2}\pi$$

$$0 \leqslant \cos^{-1} x \leqslant \pi$$

$$-\tfrac{1}{2}\pi \leqslant \tan^{-1} x < \tfrac{1}{2}\pi$$

Differentiation

f(x)	f'(x)
x^n	nx^{n-1}
$\ln x$	$\dfrac{1}{x}$
e^x	e^x
$\sin x$	$\cos x$
$\cos x$	$-\sin x$
$\tan x$	$\sec^2 x$
uv	$u\dfrac{dv}{dx} + v\dfrac{du}{dx}$
$\dfrac{u}{v}$	$\dfrac{v\dfrac{du}{dx} - u\dfrac{dv}{dx}}{v^2}$

If $x = f(t)$ and $y = g(t)$ then $\dfrac{dy}{dx} = \dfrac{dy}{dt} \div \dfrac{dx}{dt}$

Integration

f(x)	$\displaystyle\int f(x)\,dx$
x^n	$\dfrac{x^{n+1}}{n+1} + c \ (n \neq -1)$
$\dfrac{1}{x}$	$\ln|x| + c$
e^x	$e^x + c$
$\sin x$	$-\cos x + c$
$\cos x$	$\sin x + c$
$\sec^2 x$	$\tan x + c$

$$\int u\frac{dv}{dx}\,dx = uv - \int v\frac{du}{dx}\,dx$$

$$\int \frac{f'(x)}{f(x)}\,dx = \ln|f(x)| + c$$

Vectors

If $\mathbf{a} = a_1\mathbf{i} + a_2\mathbf{j} + a_3\mathbf{k}$ and $\mathbf{b} = b_1\mathbf{i} + b_2\mathbf{j} + b_3\mathbf{k}$ then

$$\mathbf{a}.\mathbf{b} = a_1 b_1 + a_2 b_2 + a_3 b_3 = |\mathbf{a}||\mathbf{b}| \cos \theta$$

Numerical integration

Trapezium rule:

$$\int_a^b f(x)\,dx \approx \tfrac{1}{2}h\{y_0 + 2(y_1 + y_2 + \ldots + y_{n-1}) + yn\}, \text{ where } h = \frac{b-a}{n}$$

Sample examination papers

Time allowed 1 hour 45 minutes

*Answer **all** the questions. Give non-exact numerical answers correct to 3 significant figures, or 1 decimal place in the case of angles in degrees, unless a different level of accuracy is specified in the question. The use of an electronic calculator is expected, where appropriate. You are reminded of the need for clear presentation in your answers.*

The number of marks is given in brackets [ ] at the end of each question or part question.

The total number of marks for this paper is 50.

Note: The number of marks for each question reflects the amount of working required in the answer.

Sample paper 1

Q1 The function f is defined by

$$f(x) = x^2 \qquad \text{for all real values of } x,$$

and the function g is defined by

$$g(x) = 4 - x \qquad \text{for all real values of } x.$$

(i) Find

    (a) $fg(x)$ (b) $gf(x)$ [2]

(ii) State the range of each of (a) $fg(x)$ and (b) $gf(x)$. [2]

Q2 The diagram shows points A and B on the circumference of a circle with centre O and radius 4 cm. The angle AOB is θ radians. The area of the sector AOB is 12 cm². Find

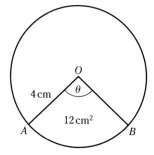

(i) θ [2]

(ii) the perimeter of the sector AOB. [2]

Q3 Solve the equations

$$y = 3 - 2x$$
$$x^2 + xy + y^2 = 3$$

[5]

Q4 (i) The second term of an arithmetical progression is 30 and the fifth term is 15. Find the common difference and the first term. [3]

(ii) The second term of a geometrical progression is 10 and the sum to infinity is 45. Find the two possible values of the common ratio. [4]

Q5 The points A, B, C have coordinates $(-2, 3)$, $(1, -4)$, $(8, -1)$ respectively.

(i) Show that AB is perpendicular to BC. [3]

(ii) Find the point at which the line through B parallel to AC meets the x-axis. [5]

Q6

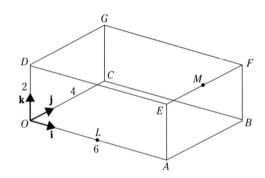

In the diagram, *OABCDEFG* is a cuboid in which *OA* = 6 units, *OC* = 4 units and *OD* = 2 units. Unit vectors **i**, **j**, **k** are parallel to $\overrightarrow{OA}$, $\overrightarrow{OC}$, $\overrightarrow{OD}$ respectively. The mid-points of *OA* and *EF* are *L* and *M* respectively.

(i) Express each of the vectors $\overrightarrow{LM}$ and $\overrightarrow{CM}$ in terms of **i**, **j**, **k**. [4]

(ii) Show that the acute angle between the directions of *LM* and *CM* is 48.8°, correct to the nearest 0.1°. [4]

Q7 (i) Solve the equation

$$2 \sin (2\theta + 10°) = 1 \text{ for } 0° \leqslant \theta \leqslant 360°$$ [5]

(ii) Prove the identity

$$\frac{\sin A}{1 + \cos A} + \frac{1 + \cos A}{\sin A} \equiv \frac{2}{\sin A}, \qquad (\cos A \neq -1).$$ [4]

Q8 A curve has equation

$$y = x + \frac{16}{x}$$

(i) Find the coordinates of the stationary points on the curve. [6]

(ii) Determine whether each of the stationary points is a maximum point or a minimum point. [3]

Q9 (i) (a) Write down the first three terms in the binomial expansion of $(1 - 2x)^8$. [3]

(b) Hence deduce the value of 0.98^8 to three decimal places. [2]

(ii) Show that

$$(x + y)^5 - (x - y)^5 = 10x^4y + 20x^2y^3 + 2y^5$$

and deduce the value of $(\sqrt{2} + 1)^5 - (\sqrt{2} - 1)^5$. [5]

Q10 A curve has equation

$$y = 1 + \frac{9}{x^2}$$

(i) Find the area of the region bounded by the curve, the lines $x = 1$, $x = 3$ and the *x*-axis. [4]

(ii) Find the volume generated when the area found in part (i) is rotated through 360° about the *x*-axis. [7]

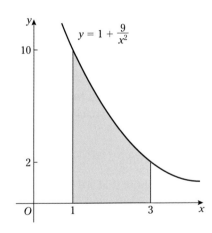

Sample paper 2

Q1 A sector with an angle of $\frac{3}{8}\pi$ radians is cut from a circle of radius 12 cm. Calculate

    (i) the perimeter of the sector   [2]

    (ii) the area of the sector.   [2]

    (Leave π in your answers.)

Q2 (i) The fourth term of an arithmetic progression is 21 and the common difference is -3. Find the first term and the sum of all the positive terms.   [3]

    (ii) The sum to infinity of a geometric progression is five times the first term. Find the common ratio.   [2]

Q3 The coordinates of the points A, B, C are $(-1, 2)$, $(5, 1)$, $(3, -4)$ respectively. Find the equation of the line through B and the mid-point of AC in the form $ax + by = c$, where a, b, c are constants to be found.   [5]

Q4 Solve the equation

$$x^6 - 7x^3 - 8 = 0$$ [6]

Q5 (i) Sketch the graph of the curve with equation $y = 2\cos(x - 45°) = -1$ for $0° \leqslant x \leqslant 360°$ [3]

    (ii) Solve the equation

$$2\cos(x - 45°) = -1 \quad \text{for} \quad 0° \leqslant x \leqslant 360°$$ [4]

Q6 A spherical balloon is being inflated at a steady rate of 8 cm³s⁻¹. Find the rate of increase of the surface area of the balloon when its radius is 4 cm.

(For a sphere: Volume $\frac{4}{3}\pi r^3$,   Surface area $= 4\pi r^2$).   [7]

Q7 The function f is defined by

    $f(x) = 4x$ for all real values of x,

and the function g is defined by

    $g(x) = \frac{1}{2}x + 1$ for all real values of x.

Find the inverse functions f^{-1} and g^{-1} and verify that $(fg)^{-1} = g^{-1}f^{-1}$ [8]

Q8 The diagram shows a curve C with equation

    $y = 5 + 4x - x^2$

and a line l with equation

    $y = 9 - x$

    (i) Find the x-coordinates for the intersections of C and l.   [4]

    (ii) Find the area of the region enclosed between C and l.   [6]

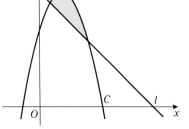

Q9 The points A and B have position vectors $3\mathbf{i} - 4\mathbf{j}$ and $\mathbf{i} - 2\mathbf{j} + 2\mathbf{k}$ respectively, relative to an origin O.

    (i) Find the lengths of OA and OB.   [3]

    (ii) Find the scalar product of $\overrightarrow{OA}$ and $\overrightarrow{OB}$, and find the angle AOB.   [4]

    (iii) Find the position vector of M, the mid-point of AB and verify that

$$OA^2 + OB^2 = 2(AM^2 + OM^2)$$ [4]

Q10   (i) Find the coordinates of stationary points on the curve with equation

$$y = 3x^2 + 12x - 2x^3 + 1$$ [7]

    (ii) Determine the nature of the stationary points.   [3]

    (iii) Sketch the curve.   [3]

P1 Answers

Numerical answers that are not exact are given correct to 3 significant figures or 1 decimal place in the case of angles in degrees, unless a different level of accuracy is specified in the question.

Chapter 1

Exercise 1a

1. $x = -2$ or $x = -3$
2. $x = 2$ or $x = -3$
3. $x = 3$ or $x = -2$
4. $x = -2$ or $x = -4$
5. $x = 1$ or $x = 3$
6. $x = 1$ or $x = -3$
7. $x = -1$ or $x = -\frac{1}{2}$
8. $x = 2$ or $x = \frac{1}{4}$
9. $x = 1$ or $x = -5$
10. $x = 8$ or $x = -9$
11. $-1, 3$
12. $-1, -4$
13. $1, 5$
14. $2, -5$
15. $-2, 7$
16. $2, 7$

Exercise 1b

1. $x = 2$ or $x = 5$
2. $x = 3$ or $x = -5$
3. $x = 4$ or $x = -1$
4. $x = 3$ or $x = 4$
5. $x = \frac{1}{3}$ or $x = -1$
6. $x = -1$ or $x = -6$
7. $x = 0$ or $x = 2$
8. $x = -1$ or $x = -\frac{1}{4}$
9. $x = \frac{2}{3}$ or $x = -1$
10. $x = 0$ or $x = -\frac{1}{2}$
11. $x = 0$ or $x = -6$
12. $x = 0$ or $x = 10$
13. $x = 0$ or $x = \frac{1}{2}$
14. $x = 5$ or $x = -4$
15. $x = 2$ or $x = -\frac{4}{3}$
16. $x = 2$ or $x = -1$
17. $x = 0$ or $x = 1$
18. $x = 0$ or $x = 2$
19. $x = 3$ or $x = -1$
20. $x = -1$ or $x = \frac{1}{2}$

Exercise 1c

1. 4
2. 1
3. 9
4. 25
5. 2
6. $\frac{25}{4}$
7. 192
8. 81
9. 200
10. $\frac{1}{4}$
11. $\frac{1}{3}$
12. $\frac{9}{8}$
13. $x = -8.12$ or 0.123
14. $x = -0.732$ or 2.73
15. $x = -1.62$ or 0.618
16. $x = 0.366$ or -1.37
17. $x = -2.62$ or -0.382
18. $x = 1.28$ or -0.781
19. $x = 0.449$ or -4.45
20. $x = -0.768$ or 0.434
21. $x = 1.12$ or -3.12
22. $x = 2.30$ or -1.30
23. $x = -0.640$ or 0.390
24. $x = 2.35$ or -0.851

Exercise 1d

1. $x = -3.41$ or -0.586
2. $x = -0.781$ or 1.28
3. $x = -4.79$ or -0.209
4. $x = 1.69$ or -1.19
5. $x = 3.73$ or 0.268
6. $x = 1.85$ or -1.35
7. $x = 0.768$ or -0.434
8. $x = -0.768$ or 0.434
9. $x = -0.260$ or -1.540
10. $x = 2.781$ or 0.719
11. $x = 1.883$ or -0.133
12. $x = 0.804$ or -1.554
13. $x = 0.804$ or -1.554
14. $x = 0.724$ or 0.276
15. $x = 7.873$ or 0.127
16. $x = 3.303$ or -0.303

Exercise 1e

1. 2 or -2
2. -1.73 or 1.73
3. 1.41 or -1.41
4. $0, 1.802$ or -1.802
5. 1.5 or -1.5
6. $0, 0.707$ or -0.707
7. 0.167
8. 0.577 or -0.577

9 0.707, −0.707, 0.447 or −0.447

10 1 or −1

11 2.45, −2.45, 1.414, −1.414

12 2.65, −2.65, 1 or −1

13 1 or 4

14 $\frac{9}{4}$

15 2.049, −2.049, 1.94 or −1.94

16 There are no real values of x as x^2 is negative.

Exercise 1f

1 4

2 $-\frac{5}{3}$

3 1

4 $\frac{4}{3}$

5 −3

6 $\frac{2}{5}$

7 real and different

8 not real

9 real and different

10 real and equal

11 real and different

12 real and equal

13 real and different

14 not real

15 real and different

16 real and equal

17 $k = \pm 12$

18 $a = 2\frac{1}{4}$

19 $p - 2$

22 $q^2 = 4p$

Exercise 1g

1
x	−2	1
y	−1	2

2 $x = -1, y = 3$

3 $x = 2, y = 3$

4
x	−1	$\frac{1}{2}$
y	4	1

5
x	2	$-\frac{1}{2}$
y	−3	2

6
x	$\frac{7}{2}$	−2
y	$-\frac{1}{2}$	5

7
x	1	2
y	2	1

8
x	−1	3
y	−4	4

9 $x = 1, y = 5$

10
x	6	−6
y	2	−4

11 $x = \frac{1}{2}, y = -1$

12
x	1	0
y	$\frac{1}{3}$	$\frac{2}{3}$

13 $x = -1, y = -\frac{1}{2}$

14 $x = 1, y = -\frac{1}{3}$

15
x	$-\frac{1}{3}$	$\frac{2}{3}$
y	$-\frac{1}{2}$	$\frac{1}{4}$

16 $x = 1, y = \frac{1}{2}$

17
x	−3	6
y	−3	$\frac{3}{2}$

18
x	1	$-\frac{1}{4}$
y	−2	3

19
x	−1	2
y	$\frac{1}{3}$	$-\frac{1}{6}$

20
x	$\frac{1}{2}$	0
y	$\frac{1}{2}$	1

21
x	1	$3\frac{1}{2}$
y	1	−4

22
x	−1	$7\frac{1}{2}$
y	2	$-\frac{7}{5}$

Mixed exercise 1

1 −1, 6

2 6.74 or −0.742

3 0.281 or −1.78

4 0.786 or −2.12

5 1, 1

6 $-\frac{1}{4}$, 3

7 1.30 or −2.30

8 −6, 2

9 0.732 or −2.73

10 −2, −2

11 0, 2

12 8.22 or −1.22

13 0

14 1 or −1

15 1.88

16 $x = \pm 2.24$ or ± 1.73

17 $x = 4, y = 5$ or $x = -22, y = 31$

18 $x = 2, y = 5$ or $x = 4, y = 9$

19 (a) 6; 4, 2   (b) $-\frac{5}{4}; -\frac{5}{8} \pm \frac{\sqrt{73}}{8}$

20 (a) not real (b) real and different
(c) real and equal (d) real and different
21 4, −1
23 2

Chapter 2
Exercise 2a

1 (a) 5 (b) 1.41 (c) 3.61
2 (a) $\left(\frac{5}{2}, 4\right)$ (b) $\left(\frac{5}{2}, \frac{1}{2}\right)$ (c) $\left(3, \frac{7}{2}\right)$
3 (a) 10.4, $\left(\frac{1}{2}, 1\right)$
(b) 2.24, $\left(-\frac{1}{2}, -1\right)$
(c) 2.83, (−2, −3)
4 8.06
5 3.61
6 (2, −4)
8 (b) $\left(-3\frac{1}{2}, -\frac{1}{2}\right)$ (c) $17\frac{1}{2}$ square units
9 (a) 7.63 (b) $\left(0, 4\frac{1}{2}\right)$
(c) $2\frac{1}{2}$
11 (−5, −3)

Exercise 2b

1 (a) 3 (b) $\frac{3}{2}$ (c) $\frac{1}{3}$
(d) $\frac{3}{4}$ (e) −4 (f) 6
(g) $-\frac{7}{3}$ (h) $-\frac{3}{2}$ (i) $\frac{k}{h}$
2 (a) yes (b) no
(c) yes (d) yes
3 (a) parallel (b) perpendicular
(c) perpendicular (d) neither
(e) parallel

Exercise 2c

1 $a = 0, b = 4$
2 (b) $22\frac{1}{2}$ square units
5 $\sqrt{a^2 + 4b^2}$
8 $\left(\dfrac{p+q}{2}, \dfrac{p+q}{2}\right)$
9 $(a − 2)^2 + (b − 1)^2 = 9$
10 8
11 $b(d − b) = ac$
12 $b^2 = 8a − 16$

Chapter 3
Exercise 3a

1 (a) $y = 2x$ (b) $y + 2x = 0$
(c) $3y = x$ (d) $4y + x = 0$
(e) $y = 0$ (f) $x = 0$
2 (a) $2y + x = 0$ (b) $2x − 3y = 0$
(c) $2x + y = 0$
3 (a) $5x − y − 17 = 0$ (b) $x + 7y + 11 = 0$

Exercise 3b

1 (a) $y = 3x − 3$
(b) $5x + y − 6 = 0$
(c) $x − 4y − 4 = 0$
(d) $y = 5$
(e) $2x + 5y − 21 = 0$
(f) $15x + 40y + 34 = 0$
2 (a) $3x − 2y + 2 = 0$
(b) $3x − 2y + 7 = 0$
(c) $x = 3$
3 **a**, **c** and **d**
4 $x + y − 7 = 0$
5 (a) $x + 2y − 5 = 0$
(b) $16x − 6y + 19 = 0$
(c) $10x − 16y + 23 = 0$
6 $2x + y = 0$
7 $4x − 5y = 0$
8 $5x − 4y = 0$
9 $x + 2y − 11 = 0$
10 $3x − 4y + 19 = 0$

Exercise 3c

2 $\frac{1}{4}$ square units
3 $\left(-\frac{4}{3}, \frac{11}{3}\right)$, (−5, 0), (6, 0)
4 $10x − 26y − 1 = 0$
5 $x + 3y − 11 = 0, \left(\frac{13}{5}, \frac{14}{5}\right)$
6 $y = 2x − 3$
7 (a) $\sqrt{20}$ (b) $x − 2y + 1 = 0$
8 $\left(\frac{9}{10}, \frac{17}{10}\right)$ and $\left(-\frac{18}{10}, \frac{26}{10}\right)$, $\left(-\frac{27}{10}, -\frac{1}{10}\right)$ or $\left(\frac{36}{10}, \frac{8}{10}\right)$,
$\left(\frac{27}{10}, -\frac{19}{10}\right)$
9 (a) (1, 2), (5, 2), (3, 6) (b) 8

Chapter 4
Exercise 4a

1 (a) $\frac{\pi}{4}$ (b) $\frac{5\pi}{6}$ (c) $\frac{\pi}{6}$
(d) $\frac{\pi}{2}$ (e) $\frac{3\pi}{2}$ (f) $\frac{2\pi}{3}$
2 (a) 30° (b) 180° (c) 18°
(d) 60° (e) 150° (f) 15°
(g) 210° (h) 135° (i) 20°
(j) 270° (k) 80° (l) 45°
(m) 108° (n) 22.5°
3 (a) 0.61 (b) 0.82 (c) 1.62
(d) 4.07 (e) 0.25 (f) 2.04
(g) 6.46
4 (a) 97.4° (b) 190.2° (c) 57.3°
(d) 119.7° (e) 286.5° (f) 360.0°
5 (a) 0.932 (b) 0.939
(c) 9.89 (d) −0.801

Exercise 4b

1 $\frac{2\pi}{3}$ cm
2 $\frac{25\pi}{2}$ cm
3 2.4 rad
4 0.692 rad
5 $\frac{15}{\pi}$ cm
6 $\frac{25}{\pi}$ cm
7 4π cm
8 $\frac{60}{\pi}$ cm
9 7.5 cm
10 85.6 cm

Exercise 4c

1 4.19 cm²
2 75.4 cm²
3 $\frac{\pi}{2}$
4 0.96 rad
5 $\frac{125\pi}{3}$ cm²
6 $\frac{15}{\pi}$ cm, $\frac{225}{2\pi}$ cm²
7 6 cm
8 $4\sqrt{3}$ cm
9 8 cm
10 0.283 rad
11 (a) 12 cm²   (b) 23.2 cm²
12 14.5 mm², 139 mm²
13 (a) 15.2 cm   (b) 32.5 cm²
14 19.6 cm, 108 cm²

Exercise 4d

3 17.8 cm²
6 (a) $\frac{\pi}{3}$
7 (a) $\frac{(50 - 2r)}{r}$
8 0.989 cm²
9 10.2 cm
10 57.1 cm²
11 (a) 1.6 rad   (b) 53.0 cm²
12 236 cm²

Summary exercise 1

1 (i) 6.67 cm   (ii) 10.3 cm²
2 $P = 18 - 6\sqrt{3} + 2\pi$
3 (ii) 61.1 cm   (iii) 281 cm²
4 (i) $y - 3 = -3(x - 2)$
   (ii) $B(0, 9), D(4, -3)$ (iii) 40
5 (6.2, 9.6)
6 (i) $2, m = 1$ (ii) $C(-1, 6)$ (iii) $D(5, 12)$
7 (ii) 58.0 cm²
8 $B(6, 5), C(12, 8)$
9 (i) 25.9 cm   (ii) 15.3 cm²
10 (i) (4, 6)   (ii) (6, 10)   (iii) 40.9
11 (i) $4\sqrt{3}$ (ii) $48\sqrt{3} - 24\pi$
12 (i) 21.5 cm²   (ii) 20.6 cm
13 (i) $(-3, 11), (2, 6)$ (ii) $y = x + 9$

Chapter 5

Exercise 5a

1 yes
2 yes
3 no, undefined when $x = 0$
4 no
5 yes, $x \geqslant 0$
6 no, undefined when $x < 0$

Exercise 5b

1 $-4, -24$
2 25, 217
3 1, not defined, 12
4 $1, \frac{\sqrt{3}}{2}$
5 (a) $f(x) \geqslant -3$ (b) $f(x) \geqslant -5$
   (c) $f(x) \geqslant 0$ (d) $0 < f(x) \leqslant \frac{1}{2}$
6 (a) 5, 4, 2, 0   (b)

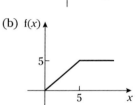

7 (a) 0, 2, 4, 5, 5   (b)

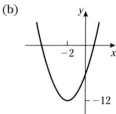

   (c) $0 \leqslant f(x) \leqslant 5$

Exercise 5c

1 (a) $\frac{11}{4}$ (b) 3   (c) 4
2 (a) $f(x) \leqslant \frac{29}{4}$ (b) $f(x) \geqslant -2$ (c) $f(x) \leqslant 1$
3 (a)

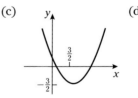

   (b)

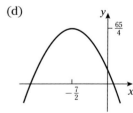

   (c)   (d)

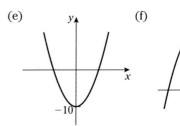

   (e)   (f)

4 (a) 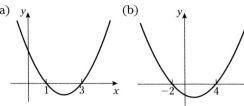 (b)

2 (a)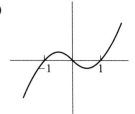

(b)

(c) 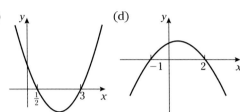 (d)

(b)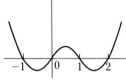

(c)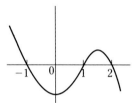

(e) 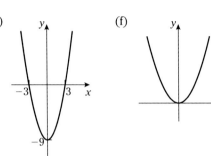 (f)

(d)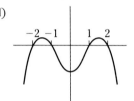

Exercise 5d

1 (a) 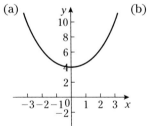 (b)

(c)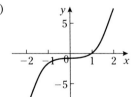

(d)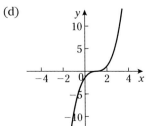

Exercise 5e

1 (a) 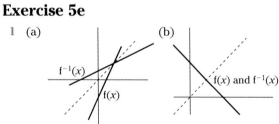 (b)

(c) (d)

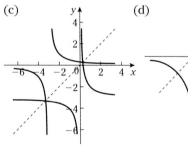

2 **b** and **d**

3 (a) $f^{-1}(x) = (x - 1)$
  (b) no
  (c) $f^{-1}(x) = \sqrt[3]{x - 1}$
  (d) $f^{-1}(x) = \sqrt{x + 4}, x \geqslant -4$
  (e) $f^{-1}(x) = \sqrt[4]{x} - 1, x \geqslant 0$

4 (a) $-\frac{1}{3}$ (b) $\frac{1}{2}$
  (c) there isn't one

Exercise 5f

1 (a) $(2x + 1)^2, x \in \mathbb{R}$ (b) $(1 - x)^2, x \in \mathbb{R}$
  (c) $-2x, x \in \mathbb{R}$ (d) $1 - x^2$
  (e) $2x^2 + 1, x \in \mathbb{R}$
2 (a) 125   (b) 15
  (c) -1 (d) -1
3 (a) $(1 + x)^2$ (b) $2(1 + x)^2$
  (c) $1 + 4x^2$
4 $g(x) = x^2, h(x) = 2 - x$
5 $g(x) = x^4, h(x) = (x + 1)$
6 (a) $f(x) = gh(x), g(x) = x^2, h(x) = 3x - 2$
  (b) $f(x) = gh(x), g(x) = x^3, h(x) = 2x + 1$
  (c) $f(x) = gh(x), g(x) = x^4, h(x) = 5x - 6$
  (d) $f(x) = g(x)h(x), g(x) = x - 1, h(x) = x^2 - 2$

Mixed exercise 5

1 (a) 16   (b)

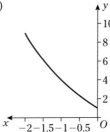

  (c) $f^{-1}(x) = 1 - \sqrt{x}$
2 (a) $\frac{11}{4}$ when $x = \frac{3}{2}$ $f(x) \geqslant 1\frac{1}{2}$
  (b) $-\frac{41}{8}$ when $x = \frac{7}{4}$ $f(x) \geqslant -\frac{41}{8}$
  (c) -9 when $x = -2$ $f(x) \geqslant -9$
3 (a) $3 - 3x$ (b) $4 - 9x$
  (c) $1 - \frac{3}{x}$ (d) $\frac{1}{3}(1 - x)$
4 (a)   (b) $4(2x - 3)^2$

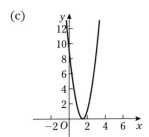

  (c)

5 (a) $f^{-1}(x) = x + 1, g^{-1}(x) = \frac{x}{1 - 2x}$
6 (a) 2   (b) $x = 2$ or 4
7 (a) 1.75   (b) 1.5

(c) $g(x)$

8 $f(x) \leqslant 0.2$
9 (a) 6   (b) $3 \leqslant f(x) \leqslant 21$

Chapter 6

Exercise 6a

1 $x < \frac{7}{2}$ 2 $x > -2$
3 $x < -\frac{1}{4}$ 4 $x > 4$
5 $x < \frac{1}{2}$ 6 $x > \frac{8}{3}$
7 $x > -3$ 8 $x < -3$
9 $x > \frac{3}{8}$

Exercise 6b

1 $x > 2$ and $x < 1$
2 $x \geqslant 5$ and $x \leqslant -3$
3 $-4 < x < 2$
4 $x \geqslant \frac{1}{2}$ and $x \leqslant -1$
5 $x > 2 + \sqrt{7}$ and $x < 2 - \sqrt{7}$
6 $-\frac{1}{2} < x < \frac{1}{2}$
7 $-4 \leqslant x \leqslant 2$
8 $x > 1$ and $x < -\frac{2}{5}$
9 $x \geqslant \frac{3}{2}$ and $x \leqslant -5$
10 $x > 4$ and $x < -2$
11 $\frac{1}{2}(-3 - \sqrt{17}) \leqslant x \leqslant \frac{1}{2}(-3 + \sqrt{17})$
12 $x > 7$ and $x < -1$
13 (a) $p \geqslant 9$ and $p \leqslant 1$
   (b) $p \geqslant 5$ and $p \leqslant 1$
14 $-2 < a < 6$
15 $p < -1, p > \frac{7}{2}$
16 $-6 \leqslant k \leqslant 6$

Exercise 6c

1 (a) $k > 0$ (b) $k > -\frac{1}{4}$
2 (a) $5 - \sqrt{20} < k < 5 + \sqrt{20}$ (b) $k = \frac{-4 \pm 4\sqrt{2}}{3}$
3 (a) -8 (b) $-\frac{1}{4}$
5 $k < 6.25$

Mixed exercise 6

1 $x < 1$
2 $x > \frac{3}{2}$

16

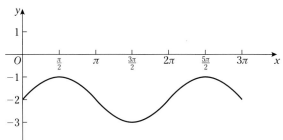

17

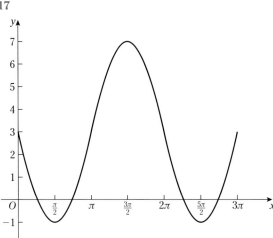

18

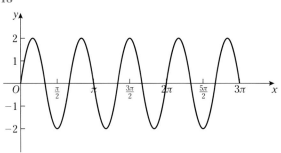

Exercise 9b

1 (a) $-\cos 57°$ (b) $-\cos 70°$

(c) $\cos 20°$ (d) $-\cos 26°$

2 (a) $-\dfrac{\sqrt{3}}{2}$ (b) 0

(c) $-\dfrac{1}{\sqrt{2}}$ (d) 1

3 (a)

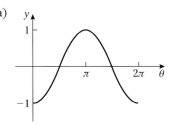

(b)

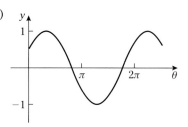

(c)

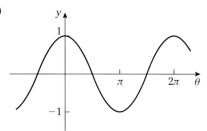

4

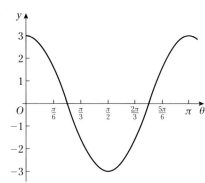

5

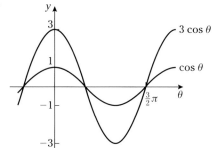

6

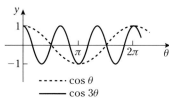

7

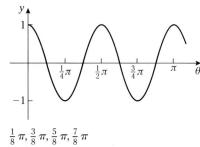

$\frac{1}{8}\pi, \frac{3}{8}\pi, \frac{5}{8}\pi, \frac{7}{8}\pi$

8 (a)
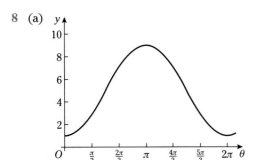

(b)
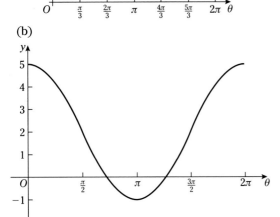

(c)
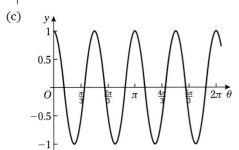

9
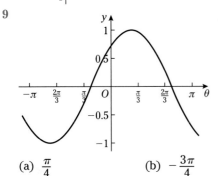

(a) $\frac{\pi}{4}$ (b) $-\frac{3\pi}{4}$

(c) $-\frac{\pi}{4}, \frac{3\pi}{4}$

Exercise 9c

1 (a) 1   (b) $-\sqrt{3}$

  (c) $\sqrt{3}$ (d) -1

2 (a) $\tan 40°$ (b) $-\tan \frac{2}{7}\pi$

  (c) $-\tan 50°$ (d) $-\tan \frac{2}{5}\pi$

3 (a) $\frac{1}{4}\pi, \frac{5}{4}\pi$ (b) $\frac{5}{4}\pi$

  (c) $0, \pi, 2\pi$ (d) $\frac{1}{2}\pi, \frac{3}{2}\pi$

4 (a)
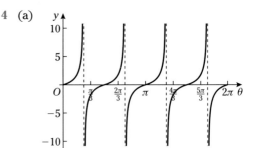

(b)
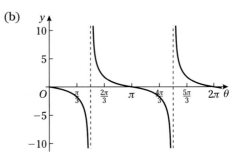

(c)
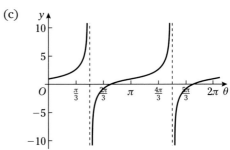

Exercise 9d

1 (a) 0.412 rad, 2.73 rad, -5.87 rad, -3.55 rad

  (b) $-\frac{4}{3}\pi, -\frac{2}{3}\pi, \frac{2}{3}\pi, \frac{4}{3}\pi$

  (c) 0.876 rad, 4.02 rad, -2.27 rad, -5.41 rad

2 (a) 141.3°, 321.3°, 501.3°, 681.3°

  (b) 191.5°, 348.5°, 551.5°, 708.5°

  (c) 84.3°, 275.7°, 444.3°, 635.7°

3 (a) 36.9°

  (b) $-36.9°$

  (c) 0.464 rad

4 $0, \pi, 2\pi$

5 11.8°, 78.2°, 191.8°, 258.2°

6 $\frac{1}{3}\pi, \pi, \frac{5}{3}\pi$

7
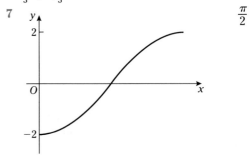

$\frac{\pi}{2}$

8

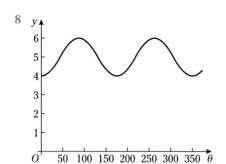

9

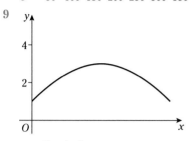

max 3, min 1

10

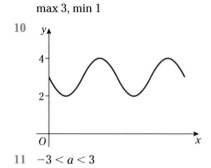

11 $-3 < a < 3$

12

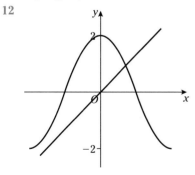

There is one solution.

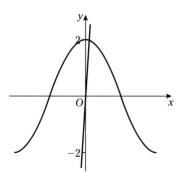

Yes, except the units for the first should be in radians and the units for the second should be in degrees.

13 (a)

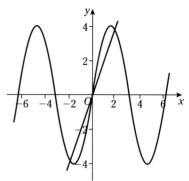

(b)

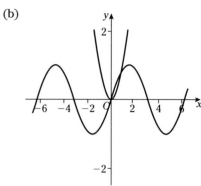

(c)

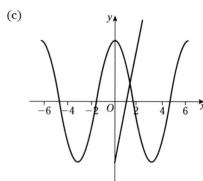

Chapter 10

Exercise 10a

1

	$\sin \theta$	$\cos \theta$	$\tan \theta$
(a)	$-\frac{12}{13}$	$-\frac{5}{13}$	$\frac{12}{5}$
(b)	$\frac{3}{5}$	$-\frac{4}{5}$	$-\frac{3}{4}$
(c)	$\frac{7}{25}$	$\frac{24}{25}$	$\frac{7}{24}$
(d)	0	± 1	0

2 $\tan^4 A$

3 1

4 $\dfrac{1}{\sin \theta \cos \theta}$

5 $\dfrac{1}{\cos^2 \theta}$

6 $\tan \theta$

7 $\sin^3 \theta$

8 $x^2 + y^2 = 16$

9 $\left(\dfrac{a^2}{x^2} + \dfrac{b^2}{y^2} \right) = 1$

10 $(x - 1)^2 + y^2 = 1$

11 $(1 - x)^2 + (y - 1)^2 = 1$

12 $x^2(b^2 - y^2) = a^2b^2$

13 $y^2(x^2 - 4x + 5) = 4$

Exercise 10b

1 $60°, 120°$

2 $90°, 270°$

3 $120°, 300°$

4 $194.5°, 345.5°$

5 $120°, 240°$

6 $45°, 225°$

7 $30°, 150°, 210°, 330°$

8 $190.1°, 349.9°$

9 $41.9°, 318.1°$

10 $-\pi, -\frac{1}{3}\pi, \frac{1}{3}\pi, \pi$

11 $-\pi, -\frac{2}{3}\pi, 0, \frac{2}{3}\pi, \pi$

12 $-\pi, -\frac{1}{6}\pi, 0, \frac{1}{6}\pi, \pi$

Exercise 10c

1 $22.5°, 112.5°$

2 $40°, 80°, 160°$

3 none

4 $67.5°, 157.5°$

5 none

6 $60°$

7 $-149.5°, -59.5°, 30.5°, 120.5°$

8 $-105.2°, -74.8°, 14.8°, 45.2°, 134.8°, 165.2°$

9 $\pm 63.6°$

10 $\frac{1}{6}\pi, \frac{5}{12}\pi, \frac{2}{3}\pi, \frac{11}{12}\pi$

11 $\frac{1}{12}\pi$

12 $\frac{1}{24}\pi, \frac{13}{24}\pi$

Exercise 10d

1 $\dfrac{\pi}{3}$

2 $-\dfrac{\pi^2}{12}$

3 $\dfrac{\pi}{2}$

4 $-\dfrac{\pi}{3}$

5 $-\dfrac{\pi}{3}$

6 $-\dfrac{\pi}{4}$

7 $\dfrac{\pi}{4}$

8 $\dfrac{\pi}{4}$

9 $\dfrac{\pi}{4}$

10 (a)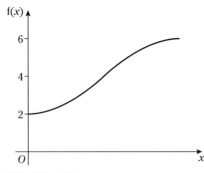

(b) yes, one−one

11 (a)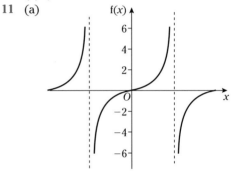

(b) no, not one−one

(c) $-\pi \leqslant x \leqslant \pi$

Mixed exercise 10

1 $x^2 + \dfrac{1}{y^2} = 1$

2 $\sin \beta = \pm \dfrac{\sqrt{3}}{2}, \tan \beta = \pm\sqrt{3}$

3 $\dfrac{2}{\sin^2 \theta}; \frac{1}{4}\pi, \frac{3}{4}\pi, \frac{5}{4}\pi, \frac{7}{4}\pi$

4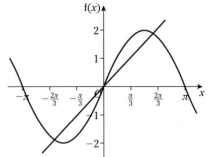

3

5 $\dfrac{\pi + 4}{12}$

6 $\frac{-29}{36}\pi, \frac{-17}{36}\pi, \frac{-5}{36}\pi, \frac{7}{36}\pi, \frac{19}{36}\pi, \frac{31}{36}\pi$

7 $(x - 2)^2 + (y + 1)^2 = 1$

8 (a) π \qquad\qquad (b) $\dfrac{\pi}{2}$

10 $\sin^2 A$

11 $\pm 70.5°, \pm 180°$

12 $a = 5, b = 4, c = 3$

13 $10°, 110°, 130°$

14 $-90°, 30°, 150°$

15 $\frac{1}{3}\pi, \frac{5}{6}\pi$

16 (a)

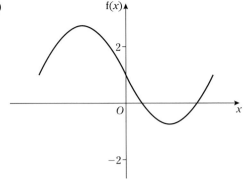

(b) e.g. $-\pi \leqslant x \leqslant \pi$

Chapter 11

Exercise 11a

1 (a) $17, 3n + 2$ (b) $-2, 8 - 2n$
    (c) $p + 4q, p - q + nq$ (d) $18, 2n + 8$
    (e) $17, 4n - 3$ (f) $0, \dfrac{5 - n}{2}$
    (g) $8, 3n - 7$
2 (a) 185 (b) -30 (c) $5(2p + 9q)$
    (d) 190 (e) 190 (f) $-\dfrac{5}{2}$
    (g) 95
3 $a = 27.2, d = -2.4$
4 $d = 3; 30$
5 $1, \dfrac{1}{2}, 0; -8\dfrac{1}{2}$
6 (a) $28\dfrac{1}{2}$ (b) 80 (c) 80
    (d) 108 (e) 40
7 $4, 2n - 4$
9 $1, 182$
10 First term $= 1$, common difference $= 5$

Exercise 11b

1 (a) 32 (b) $\dfrac{1}{8}$ (c) 48
    (d) $\dfrac{1}{2}$ (e) $\dfrac{1}{27}$
2 (a) 189 (b) -255 (c) $2 - \left(\dfrac{1}{2}\right)^{19}$
    (d) $781/125$ (e) $341/1024$ (f) 1
3 $\dfrac{1}{2}, 2$
4 $-\dfrac{1}{2}$
5 $-\dfrac{1}{2}, 1/1024$
6 (a) $\dfrac{(x - x^{n+1})}{(1 - x)}$

    (b) $\dfrac{(x^n - 1)}{[x^{n-2}(x - 1)]}$

    (c) $\dfrac{1 - y^n}{(1 - y)}$

7 $\dfrac{8}{3}\left(1 - \left(\dfrac{1}{4}\right)^n\right)$
8 10 or 62

Exercise 11c

1 (a) 6 (b) $13\dfrac{1}{3}$
    (c) $\dfrac{5}{9}$ (d) $\dfrac{9}{4}$
2 $\dfrac{1}{2}$
3 $8, 4, 2, 1$ or $24, -12, 6, -3$
4 $\dfrac{2}{3}$
5 59048
6 $16, -8$
7 2
8 1
9 $a = 2 \pm \sqrt{2}, r = \dfrac{1}{4}(2 \pm \sqrt{2})$

Exercise 11d

1 (a) $1 + 36x + 594x^2 + 5940x^3$
    (b) $1 - 18x + 144x^2 - 672x^3$
    (c) $1024 + 5120x + 11\,520x^2 + 15\,360x^3$
    (d) $1 - \dfrac{20}{3}x + \dfrac{190}{9}x^2 - \dfrac{380}{9}x^3$
    (e) $128 - \dfrac{672}{x} + \dfrac{1512}{x^2} - \dfrac{1890}{x^3}$
    (f) $\left(\dfrac{3}{2}\right)^9 + \dfrac{3^{10}}{2^7}x + \dfrac{3^9}{8}x^2 + \dfrac{7}{2}(3^7)x^3$
2 (a) $336x^2$ (b) $-10x$
    (c) $-21\,840x^{11}$ (d) $3360p^6q^4$
    (e) $16(3a)^7b$ (f) $7920x^4$
    (g) $63x^5$ (h) $56a^3b^5$
3 (a) $1 - 8x + 27x^2$ (b) $1 + 19x + 160x^2$
    (c) $2 - 19x + 85x^2$ (d) $1 - 68x + 2136x^2$
4 $1 + 18x + 144x^2 + \ldots + 512x^9$
5 8 6 1
7 $\dfrac{4}{3}$ 8 -10
9 $-\dfrac{1}{10}$ 10 200

Chapter 12

Exercise 12a

1 $\dfrac{1}{6}x^6 + K$

2 $-\dfrac{1}{4}x^{-4} + K$

3 $\dfrac{4}{5}x^{\frac{5}{4}} + K$

4 $-\dfrac{1}{2}x^{-2} + K$

5 $-\dfrac{2}{3}x^{-\frac{3}{2}} + K$

6 $2x^{\frac{1}{2}} + K$

7 $\dfrac{1}{2}x^2 + K$

8 $\dfrac{3}{2}x^{\frac{2}{3}} + K$

9 $x + \dfrac{1}{3}x^3 + K$

10 $\dfrac{1}{2}x^2 + \dfrac{1}{3}x^3 + K$

11 $x^2 - \frac{2}{3}x^{\frac{3}{2}} + K$

12 $x - \frac{1}{x} + K$

13 $x - \frac{1}{x} + K$

14 $\frac{1}{2}x^2 + \frac{1}{3}x^3 + K$

15 $2x + \left(\frac{7}{2}\right)x^2 - 5x^3 + K$

16 $2\sqrt{x} + \frac{2}{3}x^{\frac{3}{2}} + K$

17 $-\frac{1}{2x^2} + \frac{2}{x} + K$

18 $2\sqrt{x} + \frac{2}{3}x^{\frac{3}{2}} + \left(\frac{2}{7}\right)x^{\frac{7}{2}} + K$

19 $x - x^2 + \frac{1}{3}x^3 + K$

20 $\frac{1}{2}x^2 - \frac{1}{4}x^4 + K$

21 $-\frac{1}{x} + \frac{2}{\sqrt{x}} + K$

22 $-(x + 3)^{-1}$

23 $\frac{2}{3}(1 + x)^{\frac{3}{2}}$

24 $\frac{1}{18}(3x + 1)^6$

25 $\frac{1}{25}(5x - 2)^5$

26 $\frac{1}{12}(3x + 5)^4$

27 $(2x - 9)^{\frac{1}{2}}$

28 $\frac{1}{6}(4x + 1)^{\frac{3}{2}}$

Exercise 12b

1 $y = \frac{3}{2}x - x^4 + \frac{3}{2}$

2 $y = 2x + 2x^{\frac{3}{2}} - 25$

3 $y = x - \frac{3}{x} - 3$

4 $y = x^2 + 2x^3 + 5$

5 $y = \frac{94}{3} - \frac{1}{12}(2 - 3x)^4$

Exercise 12c

1 4   2 $\frac{2}{7}(8\sqrt{2} - 1)$

3 $26\frac{2}{3}$ 4 $12\frac{2}{3}$

5 15   6 -2

7 $2\frac{1}{3}$ 8 $6\sqrt{2} - 4$

9 1   10 $1\frac{7}{8}$

Exercise 12d

Answers are in square units.

1 $5\frac{1}{3}$ 2 $12\frac{2}{3}$ 3 $2\frac{2}{3}$

4 $13\frac{1}{2}$ 5 $5\frac{1}{3}$ 6 60

7 $5\frac{1}{3}$ 8 $4\frac{7}{8}$ 9 24

Exercise 12e

1 $\frac{4}{3}$

2 $15\frac{1}{4}$

3 $\frac{1}{3}$

4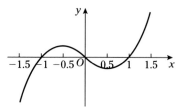

(a) $\frac{1}{4}$ (b) $\frac{1}{4}$ (c) $\frac{1}{2}$

5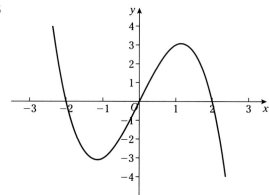

(a) 4   (b) 4   (c) 8

6 (a) -2 (b) 2   (c) 0

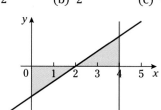

Exercise 12f

1 $\frac{1}{3}$

2 9

3 (a) $\frac{8}{3}$ (b) $\frac{16}{3}$

4 1.75

5 $\frac{1}{6}$

6 18

7 $\frac{4}{3}\sqrt{2}$

8 (a) $\frac{4}{3}$ (b) $\frac{1}{3}$

9 $4\sqrt{3}$

10 $\frac{1}{3}$

Exercise 12g

1 $\frac{1}{2}$

2 does not exist

3 $2\sqrt{2}$

4 does not exist

5 does not exist

Exercise 12h

1 $\frac{512}{15}\pi$

2 $\frac{1}{2}\pi$

3 $\frac{64}{5}\pi$

4 2π

5 8π

6 8π

7 $\frac{3}{5}\pi\left(\sqrt[3]{32}-1\right)$

8 $\frac{\pi}{2}$

9 (a) $\frac{32\pi}{5}$ (b) $\frac{8\pi}{3}$

Mixed exercise 12

1 $\frac{1}{3}x^3 + \frac{1}{x} + K$

2 $\frac{3}{4}x^{\frac{4}{3}} + K$

3 $\frac{2}{3}x^{\frac{3}{2}} + 2x^{\frac{1}{2}} + K = \frac{2}{3}\sqrt{x}\,(x+3) + K$

4 $\frac{1}{2}x^2 + \frac{1}{x} + K$

5 $\frac{2}{5}\sqrt{x}\,(x^2-5) + K$

6 9

7 $33\frac{3}{4}$

8 $\frac{3}{4}$

9 $\frac{16}{3}$

10 $\frac{9}{16}$

11 $\frac{9\pi}{14}$

12 (a) 8π (b) $\frac{256\pi}{15}$

13 $\frac{1}{2}$

Chapter 13

Exercise 13a

1 (a) $3\mathbf{i} + 6\mathbf{j} + 4\mathbf{k}$ (b) $\mathbf{i} - 2\mathbf{j} - 7\mathbf{k}$
 (c) $\mathbf{i} - 3\mathbf{k}$

2 (a) $(5, -7, 2)$ (b) $(1, 4, 0)$
 (c) $(0, 1, -1)$

3 (a) $\sqrt{21}$ (b) 5 (c) 3

4 (a) 6 (b) 7 (c) $\sqrt{206}$

5 (a) $3\mathbf{i} + 4\mathbf{k}$ (b) $2\mathbf{i} - 2\mathbf{j} + 2\mathbf{k}$
 (c) $2\mathbf{i} + 3\mathbf{j} + 3\mathbf{k}$ (d) $-6\mathbf{i} + 12\mathbf{j} - 8\mathbf{k}$

6 **b**, **d** and **e**

7 **c**

8 $\lambda = \frac{1}{2}$

9 **b**, **e** and **f**

10 (a) neither (b) parallel
 (c) equal

11 (a) $-2\mathbf{i} - 3\mathbf{j} + 7\mathbf{k}$ (b) $-\mathbf{j} + 3\mathbf{k}$
 (c) $-2\mathbf{i} - 2\mathbf{j} + 4\mathbf{k}$

12 $\sqrt{62}, \sqrt{10}, 2\sqrt{6}$

13 $\sqrt{5}$

14 $\overrightarrow{AB} = -\mathbf{i} + 3\mathbf{j}, \overrightarrow{BD} = -\mathbf{j} + 4\mathbf{k}$,
 $\overrightarrow{CD} = -2\mathbf{i} + 2\mathbf{j} - 6\mathbf{k}, \overrightarrow{AD} = -\mathbf{i} + 2\mathbf{j} - 4\mathbf{k}$

15 $\overrightarrow{OB} = 3\mathbf{i} + 3\mathbf{j}, \overrightarrow{OC} = 3\mathbf{i} + 3\mathbf{j} + 3\mathbf{k}$,
 $\overrightarrow{GF} = 3\mathbf{i} - 3\mathbf{j} + 3\mathbf{k}$

16 $\overrightarrow{OE} = 2\mathbf{i} + 3\mathbf{j}, \overrightarrow{EF} = -2\mathbf{j} + 4\mathbf{k}$,
 $\overrightarrow{FA} = -2\mathbf{i} - 3\mathbf{j} - 4\mathbf{k}$

17 $\overrightarrow{OE} = 5\mathbf{j} + 4\mathbf{k}, \overrightarrow{EB} = 3\mathbf{i} - 4\mathbf{k}, \overrightarrow{EA} = -3\mathbf{i} - 5\mathbf{j} - 4\mathbf{k}$

18 (a) $\overrightarrow{OP} = 3\mathbf{i} + 6\mathbf{j} + 6\mathbf{k}, \overrightarrow{OQ} = 6\mathbf{i} + 6\mathbf{j} + 3\mathbf{k}$,
 $\overrightarrow{PQ} = -3\mathbf{i} - 3\mathbf{k}$
 (b) 22.2 cm

Exercise 13b

1 (a) $\frac{2}{3}\mathbf{i} + \frac{2}{3}\mathbf{j} - \frac{1}{3}\mathbf{k}$ (b) $\frac{6}{7}\mathbf{i} - \frac{2}{7}\mathbf{j} - \frac{3}{7}\mathbf{k}$
 (c) $\frac{3}{5}\mathbf{i} + \frac{4}{5}\mathbf{k}$ (d) $\frac{1}{9}\mathbf{i} + \frac{8}{9}\mathbf{j} + \frac{4}{9}\mathbf{k}$
 (e) $\frac{3}{5}\mathbf{i} + \frac{4}{5}\mathbf{j}$

2 $\frac{1}{\sqrt{22}}(-2\mathbf{i} - 3\mathbf{j} + 3\mathbf{k})$

3 $\frac{1}{\sqrt{a^2 + 4a + 8}}(2\mathbf{i} - (a+2)\mathbf{j}), a = -2$

4 3

Exercise 13c

1 (a) 30
 (b) 0; **a** and **b** are perpendicular
 (c) -1

2 (a) $7, \frac{1}{3}\sqrt{7}$ (b) $14, \sqrt{\frac{7}{19}}$
 (c) $3, \frac{3}{58}\sqrt{58}$ (d) $1, \frac{1}{5}$

3 4

4 $-\sqrt{\frac{7}{34}}; 33.2°$

7 (a) $10\sqrt{3}$ (b) $\sqrt{41 - 20\sqrt{3}}$

8 $67.80°$

9 (a) (i) $4\mathbf{j} - 4\mathbf{k}$ (ii) $4\mathbf{i} + 4\mathbf{j}$
 (b) $\hat{A} = 90°, \hat{E} = 19.5°, \hat{H} = 70.5°$

10 (a) (i) $4\mathbf{k} - 4\mathbf{i}$ (ii) $-2\mathbf{i} + 4\mathbf{j} + 2\mathbf{k}$
 (b) $70.5°$

11 (b) $122°$ (c) 17.3 square units

12 (a) $17, 40.2°$
 (b) $4\mathbf{i} - 4\mathbf{j} + 3\mathbf{k}, \sqrt{206}$
 (c) $-5\mathbf{j} + \mathbf{k}, \frac{1}{3}(10\mathbf{i} - 7\mathbf{j} + 5\mathbf{k})$

13 $60°$

14 (a) $60.0°$ (b) $10.8\,\text{cm}^2$

Summary exercise 3

1 3.75

3 8

4 (i) $\overrightarrow{PA} = -6\mathbf{i} - 8\mathbf{j} - 6\mathbf{k}, \overrightarrow{PN} = 6\mathbf{i} + 2\mathbf{j} - 6\mathbf{k}$

    (ii) 99°

5 19.1° or 109.1°

6

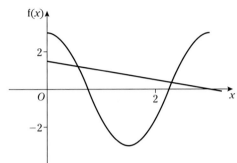

7 (i) $\overrightarrow{PR} = \begin{pmatrix} 2 \\ 2 \\ 2 \end{pmatrix} \overrightarrow{PQ} = \begin{pmatrix} -2 \\ 2 \\ 4 \end{pmatrix}$

    (ii) 61.9°   (iii) 12.8 cm

8 (i) $A(1, 5), B(4, 5), M(2, 4)$

    (ii) 18π

9 (i) $\frac{2}{3}$ (ii) 5150

10 (ii) 60° and 300°

11 (i) $a = 3$ and $b = -4$

    (ii) $x = 0.36$ and 2.78

    (iii)

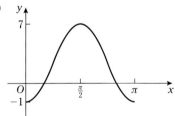

12 42π

14 (i) $\frac{1}{3}\begin{pmatrix} 1 \\ -2 \\ 2 \end{pmatrix}$ (ii) 10

    (iii) 5 or -7

Sample paper 1

1 (i) (a) $(4 - x)^2$ (b) $4 - x^2$

    (ii) (a) $fg(x) \geqslant 0$ (b) $gf(x) \leqslant 4$

2 (i) 1.5   (ii) 14 cm

3 $(1, 1), (2, -1)$

4 (i) $-5, 35$ (ii) $\frac{1}{3}, \frac{2}{3}$

5 (ii) $(-9, 0)$

6 (i) $3\mathbf{i} + 2\mathbf{j} + 2\mathbf{k}, 6\mathbf{i} - 2\mathbf{j} + 2\mathbf{k}$

7 (i) 10°, 70°, 190°, 250°

8 (i) $(-4, -8), (4, 8)$

    (ii) max, min

9 (i) (a) $1 - 16x + 112x^2$

       (b) 0.851

    (ii) 82

10 (i) 8 unit²   (ii) 40π unit³

Sample paper 2

1 (i) $24 + \frac{9}{2}\pi$ (ii) 27π cm²

2 (i) 30, 165   (ii) $\frac{4}{5}$

3 $x - 2y = 3$

4 $-1, 2$

5 (i)

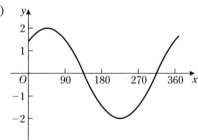

    (ii) 165°, 285°

6 4 cm²s⁻¹

7 $\frac{1}{4}x, 2x - 2$

8 (i) 1, 4   (ii) $4\frac{1}{2}$

9 (i) 5, 3

    (ii) 11, 42.8°

    (iii) $2\mathbf{i} - 3\mathbf{j} + \mathbf{k}$

10 (i) $(-1, -6), (2, 21)$ (ii) min, max

Index

A

angles 25, 79
    angle between two vectors 134
    complementary 89
    equations involving multiple 96
    radians 25–6, 27, 28–9
arcs
    length of 27–8
    and radians 25
area
    finding by definite integration
        115, 116–17, 117–19, 121–2
    sector of a circle 28–9, 30
    using integration to find 113
arithmetic progressions (APs) 100–1
    sum of an AP 101–2
asymptotes 43

B

binomial theorem 107–9

C

Cartesian coordinates 11
Cartesian unit vectors 129
    finding 133–4
    operations on 130–2
chain rule 60–1, 62
chords 54
circular measure
    area of a sector 28–9, 30
    length of an arc 27–8
    radians 25–6
common difference 100
complementary angles 89
completing the square 3–4, 5
composite functions 46–7
constant of integration 110
constants, differentiating 56–7
contact, point of 54
coordinate geometry
    Cartesian coordinates 11
    equation of a straight line 18–20
    finding the equation of a straight
        line 20–2
    gradients 13–15
    intersection 23
    length of a line joining two points
        11–12
    midpoint of a line joining two
        points 12–13
cosine function 79, 86–7, 89–90
cubic functions 42
curves
    differentiation 55–62
    equation of a line 18
    equations of tangents and normals
        65–6
    finding areas bounded by 113–22
    finding the equation of the curve
        112
    gradient 54–5
    intersection 23

intersection of a straight line and
    a curve 51–2
sketching curves of functions 40–3
stationary values 67
turning points 68–72
cyclic functions 81, 82

D

definite integration 114
    finding area by 115–22
derivative of a function 55, 110
derivatives of a function 55
differentiation 55
    the chain rule 60–1, 62
    differentiating constants and
        multiples of x 56–7
    differentiating x^n with respect
        to x 56
    equations of tangents and
        normals 65–6
    gradients of tangents and
        normals 57–8
    increasing and decreasing
        functions 59
    reversed 110
discriminants 7
displacement vectors 128, 129
domains 38–9, 39–40

E

elements 113
equal vectors 127, 130
equations 1
    equation of a normal 65
    equation of a straight line
        18–20, 20–2, 51–2
    equation of a tangent 65–6
    finding the equation of the curve
        112
    involving multiple angles 96
    quadratic 1–9
    solving using Pythagorean
        identities 94–5
expressions 1

F

factorising of quadratic equations 1–3
formulae 143–4
functions 38
    composite 46–7
    cubic 42–3
    domain of a function 38–9
    gradient (derived) function 55, 57
    increasing and decreasing 59
    integrating 110–12
    inverse 44–6
    inverse trigonometric 97–8
    polynomial 42–3
    quadratic 40–1
    range of the function 39–40
    rational 43
    trigonomic 79–91

G

geometric progressions 103
    sum of the first n terms of a GP
        104–5
    sum to infinity of a GP 105–6
gradient function 55, 57
gradients 54
    curves 54–5
    of straight lines 13–15, 16, 54
    of tangents and normals 57–8

I

identities, trigonomic 92–3, 94–5
improper integrals 121–2
inequalities 49
    manipulating 49
    solving linear 49–50
    solving quadratic 50–1
integration 110
    constant of 110, 112, 114
    definite 114
    finding an area using 113
    finding area by definite integration
        115–21
    integrating a sum or difference of
        functions 111
    volume of revolution 122–4
intersection
    of any curve and a straight line 52
    of lines or curves 23
    of a straight line and a parabola
        51–2
inverse functions 44–6
inverse trigonomic functions 97–8

L

linear inequalities 49–50
lines
    chords 54
    equation of a straight line
        18–20, 20–2
    gradient of a straight line 13–15,
        16, 54
    intersection 23, 51–2
    length of a line joining two points
        11–12
    midpoint of a line joining two
        points 12–13
    normals 54
    tangents 54
    see also curves; vectors

M

mappings 37, 38
    domain and range 38–40
maximum points 68, 69, 70
minimum points 68, 69, 70
modulus of a vector 127, 130

N

negative vectors 128
normals 54

equation of a normal 65
gradients of 57–8

O
origin (O) 11

P
parabolas 41
    intersection of a line and a
        parabola 51–2
parallel lines, gradient of 15
parallel vectors 130, 135
parameters 93
Pascal's Triangle 107
periodic functions 81, 82
perpendicular lines
    equation of 22
    gradient of 15, 16
perpendicular vectors 135
phase shifts 86
point of contact 54
point of intersection 23
polynomial functions 42–3
position vectors 129
progressions 100
    arithmetic 100–2
    binomial theorem 107–9
    geometric 103–5, 105–6
    Pascal's Triangle 107
    series 106
Pythagoras' theorem 11
Pythagorean identities 92–3, 94–5

Q
quadratic equations 1
    completing the square 3–4
    equations that are quadratic in a
        function of x 6
    formula for solving 5
    properties of the roots of 6–8
    solution by factorising 1–3
    solution of one linear and one
        quadratic equation 9
quadratic functions 40–1
quadratic inequalities 50–1

R
radians 25–6, 27, 79
range of the function 39
rational functions 43
real roots 7–8
revolution, volume of 122
roots of a quadratic equation 1
    finding 1–5
    properties of 6–8
rotation 79

S
scalar product of two vectors 134–6
scalar quantities 127
sectors of a circle, area of 28–9
sequences *see* progressions
series 106
simultaneous equations, solution
        where one is linear and the other is
        quadratic 9
sine function 79, 80–5, 89–90
sine waves 81, 82, 86
solids of revolution 122
square, completing the 3–4, 5
stationary points 67, 68, 69
stationary values 67
straight lines
    defining 18
    equation of a straight line 18–20
    finding the equation of a straight
        line 20–2
    gradients of 13–15, 16, 54
    intersection of a straight line and
        a parabola 51–2
    length of a line joining two points
        11–12
    midpoint of a line joining two
        points 12–13

T
tangent function 79, 87–8, 89–90
tangents 54
    equation of a tangent 65–6
    gradients of 54–5, 57–8

three dimensions, location of a point
    in 129
touch, meaning of the term in
    mathematics 54
trapezium rule 144
trigonometry
    circular measure 25–9
    cosine function 86–7
    equations involving multiple angles
        96
    inverse trigonometric functions
        97–8
    Pythagorean identities 92–3, 94–5
    relationships between sine, cosine
        and tangent functions 89–91
    sine function 80–5
    tangent function 87–8
    trigonometric functions 79
    trigonometric identities 92
    trigonomic ratios of 30°, 40°, 60° 80
turning points 68
    distinguishing between 69–72

U
unit vectors 129
    finding 133–4
    operations on 130–2

V
vectors 127
    the angle between two 134
    Cartesian unit vectors 129–32
    finding a unit vector 133–4
    position and displacement 128–9
    properties of 127–8
    the scalar product 134–6
volume of revolution 122–4

X
x-axis 11

Y
y-axis 11